Lecture Notes in Computer Science 5599

Commenced Publication in 1973
Founding and Former Series Editors:
Gerhard Goos, Juris Hartmanis, and Jan van Leeuwen

Claudio Bettini   Sushil Jajodia
Pierangela Samarati   X. Sean Wang (Eds.)

Privacy in Location-Based Applications

Research Issues and Emerging Trends

 Springer

Volume Editors

Claudio Bettini
Università degli Studi di Milano
Dipartimento di Informatica e Comunicazione (DICO)
Via Comelico 39, 20135 Milano, Italy
E-mail: bettini@dico.unimi.it

Sushil Jajodia
George Mason University
4400 University Drive, Fairfax, VA 22030-4444, USA
E-mail: jajodia@gmu.edu

Pierangela Samarati
Università degli Studi di Milano
Dipartimento di Tecnologie dell' Informazione
Via Bramante 65, 26013 Crema, Italy
E-mail: pierangela.samarati@unimi.it

X. Sean Wang
University of Vermont, Department of Computer Science
33 Colchester Avenue, Burlington, VT 05405, USA
E-mail: xywang@cs.uvm.edu

Library of Congress Control Number: 2009931473

CR Subject Classification (1998): K.4.1, H.2.8, K.4, C.2

LNCS Sublibrary: SL 3 – Information Systems and Application, incl. Internet/Web
and HCI

ISSN   0302-9743
ISBN-10   3-642-03510-8 Springer Berlin Heidelberg New York
ISBN-13   978-3-642-03510-4 Springer Berlin Heidelberg New York

Typesetting: Camera-ready by author, data conversion by Scientific Publishing Services, Chennai, India
Printed on acid-free paper   SPIN: 12725137   06/3180   5 4 3 2 1 0

Preface

Location-based applications refer to those that use location information in a prominent manner. These applications include, but are not limited to, location-based services (LBS). Examples of LBS include the identification of resources close to the user (e.g., the closest pharmacy) and other users in proximity (e.g., friends in the same neighborhood), as well as the identification of the optimal route to reach a destination from the user's position considering traffic conditions and possibly other constraints. LBS requests are typically invoked through mobile devices that can provide location data in terms of a user's position, movement direction, and speed. Other location-based applications use similar data, possibly stored in a moving object database, to solve different kinds of optimization problems, to perform statistical analysis of specific phenomena, or to predict potentially critical situations.

Location data can be very effective for service provisioning and can enable new kinds of information services, but they may pose serious privacy threats to users. Although data security and privacy issues have been extensively investigated in several domains, the currently available techniques are not readily applicable in location-based applications. It is a unique challenge to conciliate the effectiveness of location-based applications with privacy concerns, mostly due to the semantic richness of location and time information that is necessarily connected to location-based applications. The research in this field involves aspects of spatio-temporal reasoning, query processing, system security, statistical inference, and more importantly anonymization techniques. Several research groups have been working in recent years to identify privacy attacks and defense techniques in this domain.

This book was partially born out of the First International Workshop on Privacy in Location-Based Applications (PiLBA 2008) held in Malaga, Spain, in October 2008, in conjunction with the 13th European Symposium on Research in Computer Security. The aim of the workshop was to bring together scientists from security and data management to discuss the most recent advances in the field and the most promising research directions. The discussions at the workshop greatly influenced the structure of this book that includes extended versions of selected papers from the workshop as well as specially invited ones.

The book has two objectives. The first is to provide a solid ground for researchers approaching this topic to understand current achievements through a common categorization of privacy threats and defense techniques. This objective is particularly hard to achieve considering the recent literature on privacy in LBS, since many papers are based on specific (and often implicit) assumptions. The first four chapters are important contributions toward this objective. Chapter 1 provides a general categorization of privacy threats and defenses, and then focuses on anonymity-based approaches that can take advantage of the presence of an intermediate trusted server to anonymize both single as well as sequences of requests. Chapters 2 and 3 consider protection of location data through service architectures without trusted intermediate servers; the former focuses on defenses based on spatial cloaking, fake locations, and progressive retrieval, while the latter presents defenses based on private information retrieval techniques. Chapter 4 explains privacy challenges and defense opportunities arising when LBS are deployed in the context of hybrid mobile networks, which integrate wired, wireless and cellular technologies.

The second objective of the book is to illustrate the many facets that make the study of privacy in location-based applications a particularly challenging research subject, including topics that go beyond privacy-preserving transformations of LBS requests. Chapter 5 illustrates how location data have influenced traditional access control mechanisms, both to include location as a condition for accessing resources, and as a resource to be protected from unauthorized access. Chapter 6 addresses privacy issues in the adoption of RFID tags as devices to detect the identity as well as location of moving objects. Chapter 7 surveys privacy problems and solutions in context-aware services, in which other data in addition to or joined with location data can lead to privacy violations. Chapter 8 discusses privacy issues in the emerging area of location-based applications in vehicular ad hoc networks (VANETs). Chapter 9 concludes the book by surveying the results on privacy-preserving off-line publication of location data stored in moving object databases.

Overall, the contributions included in this book represent a wide spectrum of privacy-preserving techniques dealing with privacy issues in location-based applications. It is the editors' wish that this book offers an informative and enjoyable reading for anyone interested in these issues.

This book could not have been published without the contribution of many people, including the authors with their hard work and the reviewers with their invaluable suggestions. A special acknowledgment goes to Javier Lopez for his enthusiastic support of the PiLBA workshop, to Alfred Hofmann for making this book appear in the Springer LNCS series, and to Linda Pareschi for her support in the collection and editing of the contributions.

June 2009   Claudio Bettini
                                                            Sushil Jajodia
                                                      Pierangela Samarati
                                                             X. Sean Wang

Table of Contents

Anonymity and Historical-Anonymity in Location-Based Services

Claudio Bettini[1], Sergio Mascetti[1], X. Sean Wang[2],
Dario Freni[1], and Sushil Jajodia[3]

[1] EveryWare Lab - DICo, Università degli Studi di Milano, Italy
{bettini,mascetti,freni}@dico.unimi.it
[2] Department of CS, University of Vermont, VT, USA
sean.wang@uvm.edu
[3] Center for Secure Information Systems, George Mason University, VA, USA
jajodia@gmu.edu

Abstract. The problem of protecting user's privacy in Location-Based Services (LBS) has been extensively studied recently and several defense techniques have been proposed. In this contribution, we first present a categorization of privacy attacks and related defenses. Then, we consider the class of defense techniques that aim at providing privacy through anonymity and in particular algorithms achieving "historical k-anonymity" in the case of the adversary obtaining a trace of requests recognized as being issued by the same (anonymous) user. Finally, we investigate the issues involved in the experimental evaluation of anonymity based defense techniques; we show that user movement simulations based on mostly random movements can lead to overestimate the privacy protection in some cases and to overprotective techniques in other cases. The above results are obtained by comparison to a more realistic simulation with an agent-based simulator, considering a specific deployment scenario.

1 Introduction

Location-based services (LBS) have recently attracted much interest from both industry and research. Currently, the most popular commercial service is probably car navigation, but many other services are being offered and more are being experimented, as less expensive location aware devices are reaching the market. Consciously or unconsciously, many users are ready to give up one more piece of their private information in order to access the new services. Many other users, however, are concerned with releasing their exact location as part of the service request or with releasing the information of having used a particular service [1]. To safeguard user privacy while rendering useful services is a critical issue on the growth path of the emerging LBS.

An obvious defense against privacy threats is to eliminate from the request any data that can directly reveal the issuer's identity, possibly using a pseudonym whenever this is required (e.g., for billing through a third party). Unfortunately,

C. Bettini et al. (Eds.): Privacy in Location-Based Applications, LNCS 5599, pp. 1–30, 2009.

simply dropping the issuer's personal identification data may not be sufficient to anonymize the request. For example, the location and time information in the request may be used, with the help of external knowledge, to restrict the possible issuer to a small group of users. This problem is well-known for the release of data in databases tables [2]. In that case, the problem is to protect the association between the identity of an individual and a tuple containing her sensitive data; the attributes whose values could possibly be used to restrict the candidate identities for a given tuple are called *quasi-identifiers* [3,4].

For some LBS, anonymity may be hard to achieve and alternative approaches have been proposed, including obfuscation of sensitive information and the use of private information retrieval (PIR) techniques. For example, sensitive service parameters (possibly including location) can be generalized, partly suppressed, transformed, or decomposed using multiple queries in order to obfuscate their real precise value, while preserving an acceptable quality of service.

While the main goal of this contribution is to illustrate anonymity-based privacy protection techniques, the first two sections are devoted to a categorization of LBS privacy attacks, and to the classification of the main proposed defense techniques, including private information obfuscation and PIR, according to the threats they have been designed for, and according to other general features. This contribution does not discuss techniques aimed to the *off-line* anonymization of sets of trajectories (as in [5]), but only on techniques that are incrementally applied to service requests at the time they are issued. In Section 4, we focus on anonymity-based approaches and we show how historical k-anonymity can be achieved when an adversary has the ability to recognize sequences of requests by the same issuer. In Section 5, we report an experimental evaluation of anonymization algorithms showing the impact of realistic user movement simulations in these evaluations. Section 6 identifies some interesting research directions, and Section 7 concludes the chapter.

2   A Classification of Attacks to LBS Privacy

There is a privacy threat whenever an adversary is able to associate the identity of a user to information that the user considers private. In the case of LBS, this *sensitive association* can be possibly derived from location-based requests issued to service providers. More precisely, the identity and the private information of a single user can be derived from requests issued by a group of users as well as from available background knowledge. Figure 1 shows a graphical representation of this general privacy threat in LBS.

A *privacy attack* is a specific method used by an adversary to obtain the sensitive association. Privacy attacks can be divided into categories mainly depending on several parameters that characterize the *adversary model*. An adversary model has three main components: a) the target private information, b) the ability to obtain the messages exchanged during service provisioning, and c) the *background knowledge* and the *inferencing abilities* available to the adversary.

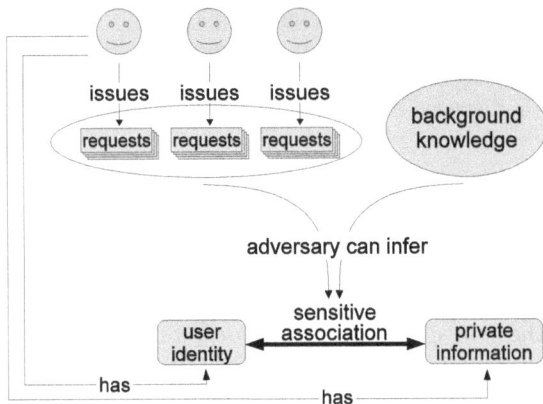

Fig. 1. General privacy threat in LBS

The target private information is the type of information that the adversary would like to associate with a specific individual, like e.g., her political orientation, or, more specifically, her location. Different classes of adversaries may also have different abilities to obtain the messages exchanged with the service provider, either by eavesdropping the communication channels or by accessing stored data at the endpoints of the communication. This determines, for example, the availability to the adversary of a single message or multiple messages, messages from a specific user or from multiple users, etc.. Finally, the adversary may have access to external knowledge, like e.g., phone directories, lists of members of certain groups, voters lists, and even presence information for certain locations, and may be able to perform inferences, like joining information from messages with external information as well as more involved reasoning. For example, even when a request does not explicitly contain the sensitive association (e.g., by using pseudo-identifiers to avoid identification of the issuer), the adversary may re-identify the issuer by joining location data in the request with presence data from external sources.

Regarding background knowledge, two extreme cases can be considered. When no background knowledge is available, a privacy threat exists if the sensitive association can be obtained only from the messages in the service protocol. When "complete" background knowledge is available, the sensitive association is included and the privacy violation occurs independently from the service request.

Hence, privacy attacks should not only be categorized in terms of the target private information, and of the availability to the adversary of service protocol messages (the first two of the main components mentioned above), but also in terms of the available background knowledge and inferencing abilities. In the following, we list some categories of privacy attacks specifically enabled by background knowledge.

- Attacks exploiting *quasi-identifiers* in requests;
- Snapshot versus historical attacks;

- Single- versus multiple-issuer attacks;
- Attacks exploiting knowledge of the defense;

Each category is discussed in the rest of this section.

2.1   Attacks Exploiting *Quasi-Identifiers*

Either part of the sensitive association can be discovered by joining information in a request with external information. When we discover the identity of the issuer (or even restrict the set of candidate issuers) we call the part of the request used in the join *quasi-identifier*. For example, when the location data in the request can be joined with publicly available presence data to identify an individual, we say that location data act as quasi-identifier. Similarly to privacy preserving database publication, the recognition of what can act as quasi-identifier in service request is essential to identify the possible attacks (as well as to design appropriate defenses).

2.2   Snapshot versus Historical Attacks

Most of the approaches presented so far in the literature [6,7,8,9] have proposed techniques to ensure a user's privacy in the case in which the adversary can acquire a single request issued by that user. More specifically, these approaches do not consider attacks based on the correlation of requests made at different time instants. An example are attacks exploiting the ability of the adversary to *link* a set of requests, i.e., to understand that the requests have been issued by the same (anonymous) user.

When historical correlation is ignored, we say that the corresponding threats are limited to the *snapshot case*. Intuitively, it is like the adversary can only obtain a snapshot of the messages being exchanged for the service at a given instant, while not having access to the complete history of messages.

In contrast with the snapshot case, in the *historical case* it is assumed that the adversary is able to *link* a set of requests. Researchers [10,11] have considered such a possibility. Several techniques exist to *link* different requests to the same user, with the most trivial ones being the observation of the same identity or pseudo-identifier in the requests, and others being based on spatiotemporal correlations. We call *request trace* a set of requests that the adversary can correctly associate to a single user. More dangerous threats can be identified in contexts characterized by the historical case as explained in [12].

2.3   Single versus Multiple-Issuer Attacks

When the adversary model limits the requests that can be obtained to those being issued by a single (anonymous) user, we say that all the attacks are *single-issuer attacks*. When the adversary model admits the possibility that multiple requests from multiple users are acquired, and the adversary is able to understand if two requests are issued by different users, we have a new important category of attacks, called *multiple-issuer attacks*. Note that this is an orthogonal classification with respect to snapshot and historical. Example 1 shows that, in the multiple-issuer case, an adversary can infer the sensitive association for a user even if the identity of that user is not revealed to the adversary.

Example 1. Suppose Alice issues a request r and that the adversary can only understand that the issuer is one of the users in a set S of potential issuers. However, if all of the users in S issue requests from which the adversary can infer the same private information inferred from r, then the adversary can associate that private information to Alice as well.

In the area of privacy in databases, this kind of attack is known as *homogeneity attack* [13]. In LBS, differently from the general case depicted in Figure 1), in the snapshot, multiple-issuer case, a single request for each user in a group is considered. More involved and dangerous threats can occur in the historical, multiple-issuer case.

2.4   Attacks Exploiting Knowledge of the Defense

In the security research area, it is frequently assumed that the adversary knows the algorithms used for protecting information, and indeed the algorithms are often released to the public. We have shown [14] that the first proposals for LBS privacy protection ignored this aspect leading to solutions subject to so called *inversion* attacks. As an example of these attacks, consider spatial cloaking as a defense technique, and suppose that a request with a certain cloaked region is observed by the adversary. Suppose also that he gets to know the identity of the four potential issuers of that request, since he knows who was in that region at the time of the request; Still he cannot identify who, among the four, is the actual issuer, since cloaking has been applied to ensure 4-anonymity. However, If he knows the cloaking algorithm, he can simulate its application to the specific location of each of the candidates, and exclude any candidate for which the resulting cloaked region is different from the one in the observed request. Some of the proposed algorithms are indeed subject to this attack. Kalnis et al. [8] show that each generalization function satisfying a property called *reciprocity* is not subject to the inversion attack. In our chapter, depending on the assumption in the adversary model about the knowledge of the defense algorithm we distinguish *def-aware attacks* from *def-unaware attacks*.

3   Defenses to LBS Privacy Threats

Defense techniques can be categorized referring to the attacks' classification reported above, depending on which specific attacks they have been designed for. However, there are other important criteria to distinguish defense approaches:

1. Defense technique: Identity anonymity versus private information obfuscation versus encryption
2. Defense architecture: Centralized versus decentralized
3. Defense validation: Theoretical versus experimental.

The different defense techniques can be classified as *anonymity-based* if they aim at protecting the association between an individual and her private information by avoiding the re-identification of the individual through a request (or a sequence of requests). This is achieved by transforming the parts of the *original request* acting as quasi-identifiers to obtain a *generalized request*. On the contrary, techniques based on *private information obfuscation* aim to protect the same association by transforming the private information contained in the original request, often assuming that the identity of the individual can be obtained. Finally, *encryption-based* techniques use private information retrieval (PIR) methods that can potentially protect both the identity of the issuer and the private information in the request.

Centralized defense architectures assume the existence of one or more trusted entities acting as a proxy for service requests and responses between the users and the service providers. The main role of the proxy is to transform requests and possibly responses according to different techniques in order to preserve the privacy of the issuers. Decentralized architectures, on the contrary do not assume intermediate entities between users and service providers. Among the benefits of centralized architectures are a) the ability of the proxy to use information about a group of users (e.g., their location) in order to more effectively preserve their privacy, and b) the availability of more computational and communication resources than the users' devices. The main drawbacks are considered the overheads in updating on the proxy the information about the users, and the need for the user to trust these entities.

A third criteria to distinguish the defenses that have been proposed is the validation method that has been used. In some cases, formal results, based on some assumptions, have been provided so that a certain privacy is guaranteed in all scenarios in which the assumptions hold. In other cases, only an experimental evaluation, usually based on synthetic data, is provided. It will be clear later in this contribution that this approach may be critical if the actual service deployment environment does not match the one used in the evaluation.

In this section we classify the main proposals appeared in the literature according to this categorization.

3.1   Anonymity Based Defenses

Most of the techniques proposed in the LBS literature to defend privacy through anonymity consider the location as a quasi-identifier. Indeed, it is implicitly or explicitly assumed that background knowledge can in some cases lead an adversary to infer the identity of the issuer given her location at a given time. Consequently, the target private information for the considered attacks is usually the specific service being requested, or the location of the issuer whenever that location cannot be used as quasi-identifier.[1]

When the location acts as a quasi-identifier, the defense technique transforms the location information in the original request into a *generalized location*. In the following we call *anonymity set* of a generalized request, the set of users that, considering location information as quasi-identifier, are not distinguishable from the issuer.

Centralized Defenses against Snapshot, Single-Issuer and Def-Unaware Attacks. Anonymity based defenses with centralized architectures assume the existence of a trusted proxy that is aware of the movements of a large number of users. We call this proxy Location-aware Trusted Server (LTS).

The first generalization algorithm that appeared in the literature is named *IntervalCloaking* [7]. The paper proposes to generalize the requests along the spatial and/or temporal dimension. For what concerns the spatial dimension, the idea of the algorithm is to iteratively divide the total region monitored by the LTS. At each iteration the current area q_{prev} is partitioned into quadrants of equal size. If less than k users are located in the quadrant q where the issuer of the request is located, then q_{prev} is returned. Otherwise, iteration continues considering q as the next area. For what concerns the temporal dimension, the idea is to first generalize the spatial location (with the above algorithm) at a resolution not finer than a given threshold. Then, the request is delayed until k users pass through the generalized spatial location. This defense algorithm has only been validated through experimental results.

An idea similar to the spatial generalization of *IntervalCloaking* is used by Mokbel et al. [9] that propose *Casper*, a framework for privacy protection that includes a generalization algorithm. The main difference with respect to *Interval-Cloaking* is that, in addition to the anonymity parameter k, the user can specify the minimum size of the area that is sent to the SP. While it is not explicit in the paper, the idea seems to be that, in addition to k-anonymity, the algorithm also provides a form of location obfuscation. Similarly to *IntervalCloaking*, *Casper* has been validated through experimental results.

Centralized Defenses against Snapshot, Single-Issuer and Def-Aware Attacks. Many papers extend *IntervalCloaking* to provide defenses techniques that guarantee anonymity when more conservative assumptions are made for the adversary model. Kalnis et al. [8], propose the *Hilbert Cloak* algorithm that provides anonymity also in the case in which the adversary knows the generalization function. The idea of *Hilbert Cloak* is to exploit the Hilbert space filling curve to define a total order among users' locations. Then, *Hilbert Cloak* partitions the users into blocks of k: the first block from the user in position 0 to the user in position $k - 1$ and so on (note that the last block can contain up to $2 \cdot k - 1$ users). The algorithm then returns the *minimum bounding rectangle* (MBR) computed considering the position of the users that are in the same block

---

[1] Indeed, location cannot be the target private information when it can be found explicitly associated with identities in background knowledge.

as the issuer. The correctness of the *Hilbert Cloak* algorithm is formally provided and the performance of the algorithm has been also experimentally evaluated.

A different algorithm, called *CliqueCloak* is proposed by Gedik et al. [15]. The main difference with respect to the *IntervalCloaking* algorithm is that *Clique-Cloak* computes the generalization among the users that actually issue a request and not among the users that are potential issuers. Indeed, *CliqueCloak* collects original requests without forwarding them to the SP until it is possible to find a spatiotemporal generalization that includes at least k pending requests. Then, the requests are generalized and forwarded to the SP. The advantage of the proposed technique, whose correctness is formally proved, is that it allows the users to personalize the degree of anonymity as well as the maximum tolerable spatial and temporal generalizations. However, the algorithm has high computational costs and it can be efficiently executed only for small values of k.

In [14] Mascetti et al. present other three generalization algorithms that are proved to guarantee anonymity against snapshot, single-issuer and def-aware attacks. The aim is to provide anonymity while minimizing the size of the generalized location. The algorithm with the best performance with respect to this metric is called *Grid*. Intuitively, this algorithm partitions all users according to their position along one dimension. Then, it considers the users in the same block as the issuer and it partitions them according to their location along the other dimension. Finally, each block has at least cardinality k and the algorithm computes the generalized location as the minimum bounding rectangle (MBR) that covers the location of the users in the same block as the issuer.

Decentralized Defenses against Snapshot, Single-Issuer Attacks. Some papers propose defense techniques that do not require a centralized architecture. Chow et al. [16] propose a decentralized solution called *CloakP2P* in which it is assumed that users can communicate with each other using an ad-hoc network. Basically, before sending the request, a user looks for the $k-1$ closest users in the neighborhood through the ad-hoc network. The location information of the request is then generalized to the region containing these users and the request is issued to the server through one of these users that is randomly selected. This algorithm guarantees privacy only against def-unaware attacks and it is evaluated through experimental results only.

Privè is a distributed protocol based on the *Hilbert Cloak* algorithm ([17]). In this case, the data structure that contains the positions of the users on the Hilbert curve is a B^+-tree that is distributed among the users in the system. The generalization is a distributed algorithm that traverses the tree starting from the root and finds the set of users containing the issuer. The algorithm is proven to be correct and guarantees privacy also against def-aware attacks. However, this solution suffers from some scalability issues. To address these issues, Ghinita et al. [18] propose the *MobiHide* algorithm which improves the scalability but that does not guarantee anonymity if the generalization algorithm is known to the adversary. The algorithm is formally validated.

A different decentralized solution is proposed by Hu et al. [19]. The main characteristic of the proposed technique is that it does not require the users to disclose their locations during the anonymization process. Indeed, it is assumed that a user's devices is able to measure the closeness from its peers through its omnidirectional antenna (using WiFi signal, for example). When a request is generalized, the distance information is used to compute the anonymity set and the generalized location is obtained through a secure computation among the users in the anonymity set. The proposed approach is safe against def-aware attacks and its correctness is formally proved.

Centralized Defenses against Historical, Single-Issuer Attacks. Several papers further extend the ideas of *IntervalCloaking* to provide a defense in the historical case. The problem of anonymity in the historical, single-issuer case has been first investigated in [12]. In the paper it is shown that the defense technique for the snapshot case cannot be straightforwardly applied to provide protection against a historical attack. In addition, a centralized algorithm is proposed. The model proposed in the paper is used in this contribution and is presented in details in Section 4.

Following the main ideas presented in [12] other anonymization techniques for the historical case have been proposed in [20,21]. The work in [20] also aims at providing protection against a def-aware attack, however it is not clear if the proposed algorithm achieves this goal since it is only evaluated through experimental results. The work in [21] proposes two generalization algorithms, the first one, called *plainKAA*, exploits the same general idea presented in [12]. The second one is an optimization of the first, based on the idea that in the generalization of the requests the users that were not in the anonymity set of a previous request can contribute to anonymity protection. It is unclear if this optimization can preserve historical k-anonymity. Both algorithms are validated through experimental results only.

Mascetti et al. propose a formal model for the historical case [22] and experimentally show that, under certain conservative assumptions, it is not possible to guarantee anonymity without generalizing the user locations to large areas. Under these assumptions, considered in most of the related work on the snaphot case, the adversary knows the association between each user identity and the location of that user. The *ProvidentHider* algorithm is proposed to guarantee anonymity in the historical case under the relaxed assumptions that the adversary knows this association only when users are located in certain areas (e.g., workplaces). The correctness of the algorithm is formally proved and its applicability is experimentally evaluated.

Centralized Defenses against Multiple-Issuer Attacks. Preliminary results on the privacy leaks determined by multiple-issuer attacks are reported in [23]. Defenses for this kind of attacks are based on accurately generalizing location (as a quasi-identifier) in order to obtain QI-groups of requests with a certain degree of *diversity* in private values. A defense against multiple-issuer attacks both in the snapshot and in a limited version of the historical case is proposed by Riboni et Al. [24] using a combination of identity anonymity and private information obfuscation techniques. Further research is needed along this line. For example, to understand under which conditions close values in private information can really be considered different (e.g., location areas).

3.2   Defenses Based on Private Information Obfuscation

As mentioned at the beginning of this section, these defenses aim at obfuscating private information released by users' requests as opposed to generalizing quasi-identifiers. To the best of our knowledge, all of the techniques in this category consider *location* as the private information to be protected, and implicitly or explicitly assume that user identity is known to the adversary or could be discovered. In the following of this chapter, we use *location obfuscation* to denote the general category of defenses aimed at obfuscating the exact location as private information of the (possibly identified) issuer.

Differently from the anonymity based defenses considering location as quasi-identifier, in this case it is less important to know the location of other users in order to provide privacy protection. For this reason, most of the location obfuscation techniques do not require a common location-aware trusted entity and, according to our categorization, they have a decentralized architecture. Sometimes these defenses are also claimed to provide a form of k-anonymity, leading to confusion with anonymity based defenses. The underlying idea is that due to the obfuscation, the location of the issuer (who is possibly not anonymous at all) cannot be distinguished among k possible locations. In order to avoid confusion this property should be called *location anonymity*.

The idea of protecting location privacy by obfuscating location information was first proposed by Gruteser et al. [25]. The technique is aimed at avoiding the association of a user with a *sensitive area* she is crossing or approaching. The proposed defense is based on appropriately suspending user requests, ensuring that the location of the user may be confused among at least other k areas. The proposed technique require a centralized entity, but it should not be difficult to modify the proposed algorithm so that it could be run directly on the users' mobile device. This defense algorithm is only validated via experiments. It is also not clear which privacy guarantees are provided if the adversary knows the algorithm.

Duckham et al. propose a protocol that allows a user to obtain the result of 1-NN (Nearest Neighbor) queries among a set of points of interest without disclosing her exact location [26]. The protocol is iterative. At the first iteration the user sends her obfuscated location to the SP that replies with the pair $\langle q, C \rangle$ where q is the point of interest having the highest confidence C of being the closest to the user. At each following iteration, the user can decide whether to provide additional location information in order to obtain a result with higher confidence. It is not specified how the generalization of the user's location is computed.

A different approach, proposed by Kido et al. [27], consists in sending, together with the real request, a set of fake requests. Since the adversary cannot distinguish the real request from the fake ones, it cannot discover the real location of the issuer, among the locations of the fake requests. This decentralized solution is effective also in the case in which the adversary knows the defense function. However, this solution has the problem that, in order to effectively protect the location information, a high number of fake requests should be sent hence impacting on the communication costs. The technique is validated through experimental results only.

In [28], Ardagna et al. propose to use a combination of location obfuscation techniques and a metric to measure the obfuscation achieved. The difference with respect to other approaches is that the resulting obfuscation area may not contain the actual location of the issuer; moreover, the location measurement error introduced by sensing technologies is taken into account. It is not formally proved that the proposed defense protects against def-aware attacks. According to our categorization, the paper considers a centralized architecture, even if the proposed obfuscation techniques can be probably run on the client side.

Recently, Yiu et al. [29] proposed a different solution to obfuscate location information, specific for LBS requests that require K-NN queries. The idea of the algorithm, named SpaceTwist, is to issue each request as if it would originate from a location different from the real user location. The request may be repeated (from the same fake location) incrementally retrieving more nearest neighbor resources, until a satisfactory answer for the real location is obtained. This solution is particularly interesting since it does not require the existence of the centralized entity that provide privacy protection and involves no range NN queries on the server side. In the paper it is also formally shown how the adversary can compute the area where the user is possibly located under the assumptions that the adversary only knows the fake location, the number of requested resources, the replies from the server and the termination condition of the algorithm.

Referring to our categorization of attacks, the existing location obfuscation defenses focus on snapshot and single-issuer attacks. Example 2 shows that, in some cases, a historical attack can further restrict the possible locations of a user.

Example 2. A request issued by Alice is obfuscated in such a way that an adversary only knows that Alice is located in an area A_1 at time t_1. After a short time, Alice issues a second request that is obfuscated in such a way that the adversary knows that Alice is located somewhere in area A_2 at time t_2. Now, assume that there is a subregion A' of A_2 such that, due to speed constraints, no matter where Alice were located in A_1 at time t_1, she has no way to get to A' at time t_2. Now the adversary knows that at time t_2, Alice cannot be located in A' and hence she must be in $A_2 \setminus A'$.

Encryption Based Defenses. We call encryption based, the defense proposals based on private information retrieval (PIR) techniques. The general objective of a PIR protocol is to allow a user to issue a query to a database without the database learning the query. In [30] this techniques is used to protect users' privacy in the LBS that computes 1-NN queries. The proposed solution is proved to solve the privacy problem under the most conservative assumptions about the adversary model as it does not reveal any information about the requests to the adversary. Nevertheless, some concerns arises about the applicability of the proposed technique. First, the proposed solution applies to 1-NN queries only and it is not clear how it could be extended to other kinds of queries like K-NN queries or range queries. Second, this technique has high computational and communication overhead. Indeed, the experimental results shown in the paper give evidence that, also using a small database of objects to be retrieved, the computation time on the server side is in the order of seconds, while the communication cost is in the order of megabytes. In particular, the amount of data that needs to be exchanged between the server and the client is larger than the size of the database itself. It is not clear for which kind of services this overhead could be tolerable.

4   Historical k-Anonymity

Most of the defenses presented in Section 3 deal with snapshot attacks, while less attention has been given to historical attacks, namely those attacks that take advantage of the acquisition of a history of requests that can be recognized as issued by the same (anonymous) user. We believe that the conditions enabling this kind of attacks are very likely to occur in LBS. In this section, we present a general algorithm for providing historical anonymity as a defense against historical attacks. Consistently with the categorization of attacks and defenses presented in Sections 2 and 3 we formally characterize the attack we are dealing with, and the proposed defense. We then present the algorithm and provide its analysis.

In the following, the format of a LBS request is represented by the triple: $\langle IdData, STData, SSData \rangle$. **IdData** may be empty, contain the identity of the issuer, or a pseudo-identifier. **STData** contains spatiotemporal information about the location of the user performing the request, and the time the request was issued. This information may be a point in 3-dimensional space (with time being the third dimension) or an uncertainty region in the same space. $STData$ is partitioned into **SData** and **TData** that contain the spatial and temporal information about the user, respectively. **SSData** contains (possibly generalized) parameters characterizing the required service and service provider. An original request is denoted with r, while the same request transformed by a defense technique is denoted with r'.

4.1   Attack Category

Before we categorize attack and defense we are interested in, we use Example 3 to show that defense techniques for the snapshot cases cannot straightforwardly be used in the historical case. This example also provides our motivation for the attack and defense categories.

Example 3. Suppose Alice requires 3-anonymity and issues a request r. An algorithm safe against def-aware attacks is used to generalize r into a request r' whose spatiotemporal region includes only Alice, Bob, and Carl. Afterwards,

Alice issues a new request r_1 that is generalized into a request r'_1 whose spatiotemporal region includes only Alice, Ann, and John. Suppose the adversary is able to link requests r' and r'_1, i.e., he is able to understand that the two requests have been issued by the same user. The adversary can observe that neither Bob nor Carl can be the issuer of r'_1, because they are not in the spatiotemporal region of r'_1; Consequently, they cannot be the issuers of r' either. Analogously, considering the spatiotemporal region in r', he can derive that Ann and John cannot be the issuers of the two request. Therefore, the adversary can identify Alice as the issuer of r' and r'_1.

In this example, in addition to adversary's ability of using location as quasi-identifier, the ability to link requests is crucial for the attack to be successful. In a general scenario, in terms of the privacy attack dimensions identified in Section 2, we deal with attacks that:

1. Exploit location and time as *quasi-identifiers* in requests, that is, the adversary can identify users by their location information;
2. Use historical request traces, that is, the adversary can link requests that have been issued by the same user;
3. Do not correlate requests or sequences of requests issued by different users. This is equivalent to consider single-issuer attacks only.
4. Exploit knowledge of the defense, that is, we assume that the adversary knows the defense algorithm.

We will formalize items 1. and 2. below in order to analyze our defense rigorously, and the remaining items are exactly as discussed in the snapshot attack cases.

Location as Quasi-Identifier. Item 1. can be formalized as follows. For users' locations, we assume that the adversary has the knowledge expressed as the following *Ident* function:

$$Ident_t : the\ Areas \longrightarrow the\ User\ sets,$$

that is, given an area A, $Ident_t(A)$ is the set of users whom, through certain means, the adversary has identified to be located in area A at time t. In the following, when no confusion arises, we omit the time instant t. We further assume that this knowledge is *correct* in the sense that these identified users in reality are indeed in area A at the time.

For a given user i, if there exists an area A such that $i \in Ident(A)$, then we say i is *identified* by the adversary. Furthermore, we say that i is *identified in* A. Note that there may be users who are also in A but the adversary does not identify them. This may happen either because the adversary is not aware of the presence of users in A, or because the adversary cannot identify these users even if he is aware of their presence. We do not distinguish these two cases as we shall see later that the distinction of the two cases does not make any perceptible difference in the ability of the adversary when the total population is large.

Clearly, in reality, there are lots of different sources of external information that can lead the adversary to estimate the location of users. Some may lead the adversary to know that a user is in a certain area, but not the exact location. For example, an adversary may know that Bob is in a pub (due to his use of a fidelity card at the pub), but may not know which room he is in. Some statistical analysis may be done to derive the *probability* that Bob is in a particular room, but this is beyond the scope of this chapter.

The most conservative assumption regarding this capability of the adversary is that $Ident(A)$ will give *exactly* all the users for each area A. It can be seen that if the privacy of the user is guaranteed in this most conservative assumption, then privacy is also guaranteed against any less precise $Ident$ function. However, this conservative assumption is unlikely true in reality, while some observed that this assumption degenerates the quality of service unnecessarily. It will be interesting to see how much privacy and quality of service change with more realistic $Ident$ functions.

Another function we assume to be known to the adversary is the following:

$$Num_t : the\ Areas \longrightarrow [0, \infty),$$

that is, given an area A, $Num_t(A)$ gives an estimate of the number of users in the area at time t. This is useful to the adversary to derive some statistical information when $Ident$ function does not recognize all the users in an area. This function can be obtained from statistical information publicly available or through some kind of counting mechanism such as tickets to a theater. Again, when no confusion arises, we do not indicate the time instant t.

Request Traces Recognized by the Adversary. In item (2) of the attack category, we assume that the adversary has the ability to link requests of the same user. This is formalized as the following function L:

$$L : the\ Requests \longrightarrow the\ Request\ sets,$$

that is, given a (generalized) request r', $L(r')$ gives a set of requests such that the adversary has concluded, through certain means, are issued by the same user who issued the request r'. In other words, all the requests in $L(r')$ are *linked* to r', although the adversary may still not know who the user is.

4.2   Defense Category

We now turn to discuss the category for our proposed defense strategy. The attacks being targeted by our defense are historical attacks more precisely described in Section 4.1. Moreover, based on the categorization of Section 3, our defense technique has the following characteristics:

1. *Defense technique:* we are using anonymity, or more specifically *historical k-anonymity*
2. *Defense architecture:* centralized; we are using LTS as our centralized defense server.

3. *Defense Validation:* we validate the effectiveness and efficiency via experiments.

As indicated in item (1) above, we use a notion of historical anonymity [12] to provide the basis for defense. To define the notion of historical anonymity, it is reasonable to assume that the LTS not only stores in its database the set of requests issued by each user, but also stores for each user the sequence of her location updates. This sequence is called *Personal History of Locations* (PHL). More formally, the PHL of user u is a sequence of 3D points $(\langle x_1, y_1, t_1 \rangle, \ldots, \langle x_m, y_m, t_m \rangle)$, where $\langle x_i, y_i \rangle$, for $i = 1, \ldots, m$, represents the position of u (in two-dimensional space) at the time instant t_i.

A PHL $(\langle x_1, y_1, t_1 \rangle, \ldots, \langle x_m, y_m, t_m \rangle)$ is defined to be *LT-consistent* with a set of requests $r_1, \ldots, r_n$ issued to a SP if for each request r_i there exists an element $\langle x_j, y_j, t_j \rangle$ in the PHL such that the area of r_i contains the location identified by the point x_j, y_j and the time interval of r_i contains the instant t_j.

Then, given the set $\bar{R}$ of all requests issued to a certain SP, a subset of requests $\bar{R}' = \{r_1, \ldots, r_m\}$ issued by the same user u is said to satisfy *Historical k-Anonymity* if there exist $k-1$ PHLs $P_1, \ldots, P_{k-1}$ for $k-1$ users different from u, such that each P_j, $j = 1, \ldots, k-1$, is LT-consistent with $\bar{R}'$.

The open problem in this case is how to generalize each request in order to obtain traces that are historical k-anonymous. One problem is that the LTS has to generalize each request when it is issued, without having the knowledge of the future users' locations nor the future requests that are to be issued. A separate problem is to avoid long traces; indeed, the longer is a trace, the more each request needs to be generalized in order to guarantee historical k-anonymity.

4.3   The *Greedy* Algorithm for Historical k-Anonymity

We now present a generalization algorithms for historical anonymity. In the next subsection we will analyze the anonymity achieved by a set of generalized requests. In the experimental section, we will present an evaluation of the effectiveness of the algorithm.

Our algorithm uses a snapshot anonymization algorithm, like *Grid*, as presented in Section 3. We modify this algorithm by adding the requirement that the perimeter of the MBR be always smaller than a user-given $maxP$ value. To achieve this, we basically recursively shrink the obtained MBR from the snapshot algorithm until its perimeter is smaller than $maxP$.

The idea of the *Greedy* algorithm was first proposed in [12] and a similar algorithm was also described in [21]. *Greedy* is aimed at preserving privacy under the attack given in Section 4.1. This algorithm computes the generalization of the first request r in a trace using an algorithm for the snapshot case. (In our implementation, we use *Grid* as the snapshot algorithm to compute the generalization of the first request.) When this first request is generalized, the set A of users located in the generalized location for the first request is stored. The generalized locations of each subsequent request r' that is linked with r is then taken as the MBR of the location of the users in A at the time of r'. As in the

---

Algorithm 1. *Greedy*

Input: a request r, an anonymity set A, anonymity level k, and a maximum perimeter $maxP$.

Output: a generalized request r' and an anonymity set A'.

Method:

1: find the MBR of all the current locations (at the time of request r) of users in A (note that if $A = \emptyset$ then the MBR is empty).
2: **if** (the perimeter of the MBR is smaller than $maxP$) **then**
3:     **if** ($|A| > 1$) **then**
4:         replace the spatial information in r with the MBR, obtaining r'
5:         let $A' = A$
6:     **else**
7:         call *Grid* algorithm* with r, k, and $maxP$, obtaining r'
8:         let A' be the set of users currently in the spatial region of r'
9:     **end if**
10: **else**
11:     recursively shrink the MBR until its perimeter is smaller than $maxP$
12:     replace the spatial region in r with the resulting MBR, obtaining r'
13:     let A' be the set of users currently located in the resulting MBR
14:     **if** ($|A'| \leq 1$) **then**
15:         call *Grid* algorithm with r, k, and $maxP$, obtaining r'
16:         let A' be the set of users currently in the spatial region of r'
17:     **end if**
18: **end if**
19: **return** r' and A'

* *Instead of* Grid, *other snapshot algorithms can be used here.*

---

modification of the *Grid* algorithm, when the MBR is smaller than $maxP$, we will recursively shrink it and exclude the users that fall out of the region. Algorithm 1 gives the pseudocode. This algorithm is called initially with the first request r and empty set $A = \emptyset$, and subsequently, it is called with the successive request and the A' returned from the previous execution.

4.4 Analysis of Anonymity

A successive use of Algorithm 1 returns a sequence of generalized requests for the user, and these generalized requests are forwarded to the SP. The question we have now is how much privacy protection such a sequence of generalized requests provides. That is, we want to find the following function:

$$Att : the\ Request\ set \times the\ Users \longrightarrow [0, 1],$$

Intuitively, given a (generalized) request r' and a user i, $Att(r', i)$ gives the probability that the adversary can derive, under the assumption of the attack category of Section 4.1, that i is the issuer of r' among all the users.

In the following of this section we show how to specify the attack function. Once the attack function is specified, we can use the following formula to evaluate the privacy value of a request:

$$Privacy(r') = 1 - Att(r', issuer(r')) \tag{1}$$

Intuitively, this value is the probability that the adversary will not associate the issuer of request r' to r'.

In order to specify the Att function, we introduce the function $Inside(i, r')$ that indicates the probability of user i to be located in $r'.Sdata$ at the time of the request. Intuitively, $Inside(i, r') = 1$ if user i is identified by the adversary as one of the users that are located in $r'.Sdata$ at time $r'.Tdata$, i.e., $i \in Ident_t(r'.Sdata)$ when $t = r'.Tdata$. On the contrary, $Inside(i, r') = 0$ if i is recognized by the adversary as one of the users located outside $r'.Sdata$ at time $r'.Tdata$, i.e., there exists an area A with $A \cap r'.Sdata = \emptyset$ such that $i \in Ident(A)$. Finally, if neither of the above cases hold, then the adversary does not know where i is. There is still a probability that i is in $r'.Sdata$. This is a much more involved case, and we first analyze the simple case, in which the adversary cannot link r' to any other requests, i.e., there is no historical information about the issuer of r'. In this case, theoretically, this probability is the number of users in $r'.Sdata$ that are not recognized by the adversary (i.e., $Num(r'.Sdata) - |Ident(r'.Sdata)|$) divided by all the users who are not recognized by the adversary anywhere (i.e., $|I| - |Ident(\Omega)|$, where I is the set of all users, and Ω is the entire area for the application). Formally,

$$Inside(i, r') = \begin{cases} 1 & \text{if } i \in Ident(r'.Sdata) \\ 0 & \text{if } \exists A : A \cap r'.Sdata = \emptyset \text{ and } i \in Ident(A) \\ \frac{Num(r'.Sdata) - |Ident(r'.Sdata)|}{|I| - |Ident(\Omega)|} & \text{otherwise} \end{cases} \tag{2}$$

Example 4. Consider the situation shown in Figure 2(a) in which there is the request r' such that, at time $r'.Tdata$, there are three users in $r'.Sdata$: one of them is identified as i_1, the other two are not identified. The adversary can also identify users i_2 and i_3 outside $r'.Sdata$ at time $r'.Tdata$. Assume that the set I contains 100 users.

Clearly, i_2 and i_3 have zero probability of being the issuers, since they are identified outside $r'.Sdata$ and due to the assumption that the spatial region of any generalized request must contain the spatial region of the original request. That is, $Inside(i_2, r') = Inside(i_3, r') = 0$. On the contrary, the adversary is

(a) First request, r'.   (b) Second request, r''.

Fig. 2. Example of attack

sure about the fact that i_1 is located in $r'.Sdata$, i.e., $Inside(i_1, r') = 1$. By Formula 2, for each user i in $I \setminus \{i_1, i_2, i_3\}$, $Inside(i, r') = 2/97$.

However, when the adversary is assumed to link r' to other requests, then we need to be more careful. We define $Inside(i, L(r'))$ to be the probability that i is located in $r.STdata$ for each request r in $L(r')$. To calculate $Inside(i, L(r'))$, we need to know the probability of a user i in area B at time t if we know that the same user was in a series areas $A_1, \ldots, A_p$ at time $t_1, \ldots, t_p$, respectively, i.e., we need estimate the conditional probability:

$$P(Inside_t(i, B) | Inside_{t_1}(i, A_1), \ldots, Inside_{t_p}(i, A_p)).$$

This conditional probability depends on many factors, including the distance between these areas and the assumed moving speed of the user. We may use historical data to study this conditional probability. Absent of the knowledge of user's moving speed or historical data, in this contribution, we use a simplifying independence assumption that the probability of a user in A is independent of where the user has been in the past. Hence, we assume

$$Inside(i, L(r')) = \Pi_{r \in L(r')} Inside(i, r),$$

where $Inside(i, r)$ is as given in Formula 2.

Example 5. Continue from Example 4 and assume a second request r'' (see Figure 2(b)) is issued after r' and that r'' is linked with r', so $L(r'') = \{r', r''\}$. We call $L(r'')$ a trace and denote it τ. At time $r''.Tdata$, there are 4 users inside $r''.Sdata$, two of which are identified as i_1 and i_2. No user is identified outside $r''.Sdata$. From the above discussion, it follows that $Inside(i_2, \tau) = Inside(i_3, \tau) = 0$ since i_2 and i_3 are identified outside the first generalized request r'. All the other users have a non-zero probability of being inside the generalized location of each request in the trace. In particular, $Inside(i_1, \tau) = 1$ since i_1 is recognized in both requests. Consider a user $i \in I \setminus \{i_1, i_2, i_3\}$. Since $Inside(i, r') = 2/97$ and $Inside(i, r'') = 2/98$, we have $Inside(i, \tau) = 0.00042$, a very small number.

Now we can obtain the attack formula:

$$Att(r', i) = \frac{Inside(i, L(r'))}{\sum_{i' \in I} Inside(i', L(r'))} \tag{3}$$

Example 6. Continue from Example 5. We now know $Att(r'', i_1) = 1/(1 + 97 * 0.00042) \approx 96\%$, $Att(r'', i_2) = Att(r'', i_3) = 0$, and $Att(r'', i) = 0.00042/(1 + 97 * 0.00042) \approx 0$ for each user i in $I \setminus \{i_1, i_2, i_3\}$.

From this example, we can observe that the independence assumption causes an overestimate of the probability of i_1 to be the issuer, but an underestimate of the probability of users other than i_1, i_2, and i_3. If we knew that a user in $r'.Sdata$ is very likely to be in $r''.Sdata$ (at the respective times), then the estimate of the attack values in Example 6 needs to be revised.

5   Impact of Realistic Simulations on the Evaluation of Anonymity-Based Defense Techniques

As we motivated in the previous sections, the correctness of an anonymity-preserving technique can be formally proved based on the specific assumptions made on the adversary model. However, in practice, different adversaries may have different background knowledge and inferencing abilities. Hence, one approach consists in stating conservative assumptions under which anonymity can be guaranteed against a broad range of potential adversaries. The drawback of this approach is clear from the conservative assumptions about location knowledge considered so far by anonymity based solutions: in order to protect from the occasional knowledge by the adversary about people present at a given location (unknown to the defender), it is (often implicitly) assumed the same knowledge for all locations. Such assumptions are not realistic and lead to overprotect the users' anonymity, hence negatively impacting on the quality of service. A different approach, taken by several researcher is experimental evaluation. Since large set of real, accurate data are very hard to obtain, in most cases experiments are based on synthetic data generated through simulators. In this section we focus on validating anonymity-based defense techniques, and we show that in order to obtain significant results, simulations must be very carefully designed. In addition to evaluating the *Greedy* algorithm as a representative of historical anonymity based defenses, we are interested in the following more general questions: a) *how much does the adversary model affect the privacy obtained by the defense according to the evaluation?*, and b) *how much does the specific service deployment model affect the results of the evaluation?*

5.1   The *MilanoByNight* Simulation

In order to carefully design the simulation, we concentrate on a specific class of LBS called *friend-finder*. A friend-finder reveals to a participating user the presence of other close-by participants belonging to a particular group (friends is only one example), possibly showing their position on a map. In particular, we consider the following service: a user issues a request specifying a threshold distance δ_A and the group of target participants (e.g., the users sharing a certain interest). The SP replies with the set of participants belonging to that group whose location is not farther than δ_A from the issuer.

A first privacy threat for a user of the friend-finder service is the association of that user's identity with the service parameters and, in particular, with the group of target participants, since this can reveal the user's interests or other private information. Even if the user's identity is not explicit in a request, an adversary can obtain this association, by using the location information of a request as a quasi-identifier.

A second privacy threat is the association of the user's identity with the locations visited by that user[2]. We recall that this association takes place

---

[2] A obfuscation-based defense against this threat, specifically designed for the friend-finder service, has recently been proposed [31].

independently from the service requests if the adversary's background location knowledge is "complete" (see Section 2). However, consider the case in which the background knowledge is "partial" i.e., it contains the association between user identity and location information only for some users in some locations at some time instants. Example 7 shows how, in this case, an adversary can exploit a set of friend-finder requests to derive location information that are not included in the background knowledge.

Example 7. User A issues a friend-finder request r_1. An adversary obtains r_1 and discovers that A is the issuer by joining the location information in the request with his background knowledge (i.e., the location information of r_1 is used as quasi-identifier). Then, A moves to a different location and issues a request r_2. The adversary obtains r_2, but in this case his background knowledge does not contain sufficient information to identify the issuer of the request. However, if the adversary can understand that r_1 and r_2 are linked (i.e., issued from the same issuer), then he derives that A is also the issuer of r_2 and hence obtains new location information about A.

We suppose that the friend-finder service is primarily used by people during entertainment hours, especially at night. Therefore, the ideal dataset for our experiments should represent movements of people on a typical Friday or Saturday night in a big city, when users tend to move to entertainment places. To our knowledge, currently there are no datasets like this publicly available, specially considering that we want to have large scale, individual, and precise location data (i.e., with the same approximation of current consumer GPS technology).

Relevant Simulation Parameters. For our experiments we want to artificially generate movements for $100,000$ users on the road network of Milan[3]. The total area of the map is 324 km^2, and the resulting average density is 308 users/km^2. The simulation includes a total of $30,000$ home buildings and $1,000$ entertainment places; the first value is strictly related to the considered number of users, while the second is based on real data from public sources which also provide the geographical distribution of the places. Our simulation starts at 7 pm and ends at 1 am. During these hours, each user moves from house to an entertainment place, spends some time in that place, and possibly moves to another entertainment place or goes back home.

All probabilities related to users' choices are modeled with probability distributions. In order to have a realistic model of these distributions, we prepared a survey to collect real users data. We are still collecting data, but the current parameters are based on interviews of more than 300 people in our target category.

Weaknesses of Mostly Random Movement Simulations. Many papers in the field of privacy preservation in LBS use artificial data generated by moving object simulators to evaluate their techniques. However, most of the simulators

---

[3] $100,000$ is an estimation of the number of people participating in the service we consider.

are usually not able to reproduce a realistic behavior of users. For example, objects generated by the Brinkhoff generator [32] cannot be aggregated in certain places (e.g., entertainment places). Indeed, once an object is instantiated, the generator chooses a random destination point on the map; after reaching the destination, the object disappears from the dataset. For the same reason, it is not possible to reproduce simple movement patterns (e.g.: a user going out from her home to another place and then coming back home), nor to simulate that a user remains for a certain time in a place.

Despite these strong limitations, we made our best effort to use the Brinkhoff simulator to generate a set of user movements with characteristics as close as possible to those described above. For example, in order to simulate entertainment places, some random points on the map, among those points on the trajectories of users, were picked. The simulation has the main purpose of understanding if testing privacy preservation over random movement simulations gives significantly different results with respect to more realistic simulations.

Generation of User Movements with a Context Simulator. In order to obtain a dataset consistent with the parameters specified above, we need a more sophisticated simulator. For our experiments, we have chosen to customize the Siafu context simulator [33]. With a context simulator it is possible to design models for agents, places and context. Therefore, it is possible to define particular places of aggregation and make users dynamically choose which place to reach and how long to stay in that place.

The most relevant parameters characterizing the agents' behavior are derived from our survey. For example, one parameter that characterizes the behavior of the agents is the average time spent in an entertainment place; This value was collected in our survey and resulted to have the following values: 9.17% of the users stays less than 1 hour, 34.20% stays between 1 and 2 hours, 32.92% stays between 2 and 3 hours, 16.04% stays between 3 and 4 hours, and 7.68% stays more than 4 hours. Details on the simulation can be found in [34].

5.2   Experimental Settings

In our experiments we used two datasets of users movements. The dataset AB (Agent-Based) was generated with the customized Siafu simulator, while the dataset MRM (Mostly Random Movement) was created with the Brinkhoff simulator. In both cases, we simulate LBS requests for the friend-finder service by choosing random users in the simulation, we compute for each request the generalization according to a given algorithm, and finally we evaluate the anonymity of the resulting request as well as the Quality of Service (QoS).

Different metrics can be defined to measure QoS for different kind of services. For instance, for the friend-finder service we are considering, it would be possible to measure how many times the generalization leads the SP to return an incorrect result i.e., the issuer is not notified of a close-by friend or, vice versa, the issuer is notified for a friend that is not close-by. While this metric is useful for this specific application, we want to measure the QoS independently from the specific kind of service. For this reason, in this chapter we evaluate how QoS degrades in terms of the perimeter of the generalized location.

In addition to the dataset of user movements, we identified other two parameters characterizing the deployment model that significantly affect the experimental results: the *number of users* in the system, which remains almost constant at each time instant and the user-required degree of indistinguishability k. These two parameters, together with the most important others, are reported in Table 1, with the values in bold denoting default values.

We also identified three relevant parameters that characterize the adversary model. The parameter P_{id-in} indicates the probability that the adversary can identify a user when she is located in a entertainment place while P_{id-out} is the probability that the adversary identifies a user in any other location (e.g., while moving from home to a entertainment place). While we also perform experiments where the two probabilities are the same, our scenario suggests as much more realistic a higher value for P_{id-in} (it is considered ten times higher than P_{id-out}). This is due to the fact that restaurants, pubs, movie theaters, and similar places are likely to have different ways to identify people (fidelity or membership cards, WiFi hotspots, cameras, credit card payments, etc.) and in several cases more than one place is owned by the same company that may have an interest in collecting data about its customers. Finally, P_{link} indicates the probability that two consecutive requests can be identified as issued by the same user.[4] While we perform our tests considering a full range of values, the specific default value reported in the table is due to a recent study on the ability of linking positions based on spatiotemporal correlation [35].

Table 1. Parameter values

Parameter	Values
dataset	**AB**, *MRM*
number of users	10k, 20k, 30k, 40k, 50k, 60k, 70k, 80k, 90k, **100k**
k	**10**, 20, 30, 40, 50, 60
P_{id-in}	0.1, **0.2**, 0.3, 0.4, 0.5, 0.6, 0.7, 0.8, 0.9, 1.0
P_{id-out}	0.01, **0.02**, 0.03, 0.04, 0.05, 0.06, 0.07, 0.08, 0.09, 0.1
P_{link}	0.1, 0.2, 0.3, 0.4, 0.5, 0.6, 0.7, 0.8, **0.87**, 0.9, 1.0

The experimental results we show in this section are obtained by running the simulation for 100 issuers and then computing the average values.

In our experiments we evaluated two generalization algorithms. One algorithm is *Greedy* which is described in Section 4 and is a representative of the historical generalization algorithm proposed so far [12,20,21]. The other algorithm is *Grid* which is briefly described in Section 3.1 is a representative of the snapshot generalization algorithms. In [14] *Grid* is shown to have better performance (in terms of the quality of service) when compared to other snapshot generalization

---

[4] The limitation to consecutive requests is because in our specific scenario we assume linking is performed mainly through spatiotemporal correlation.

algorithms like, for example, *Hilbert Cloak*. We also evaluated the privacy threat when no privacy preserving algorithm is applied. The label *NoAlg* is used in the figures to identify results in this particular case.

5.3 Impact of the Adversary Model on the Evaluation of the Generalization Algorithms

We now present a set of experiments aimed at evaluating the impact of the adversary model on the anonymity provided by the generalization algorithms.

Two main parameters characterizing the adversary model are P_{id-in} and P_{link}. In Figure 3(a) we show the average anonymity for different values of P_{id-in} when, in each test, P_{id-out} is set to $P_{id-in}/10$. As expected, considering a trace of requests, the higher is the probability of identifying users in one or more of the regions from which the requests in the trace were performed, the smaller is the level of anonymity.

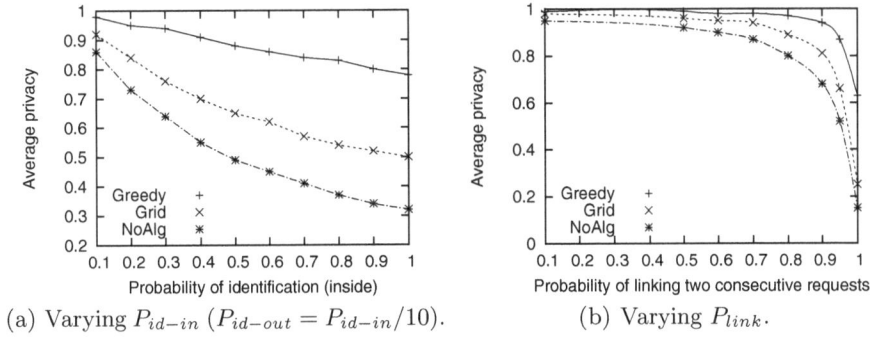

(a) Varying P_{id-in} ($P_{id-out} = P_{id-in}/10$).   (b) Varying P_{link}.

Fig. 3. Average anonymity

Figure 3(b) shows the impact of P_{link} on the average privacy. As expected, high values of P_{link} lead to small values of privacy. Our results show that the relation between the P_{link} and privacy is not linear. Indeed, privacy depends almost linearly on the average length of the traces identified by the adversary. In turn, the average length of the traces grows almost exponentially with the value of P_{link}.

To summarize the first set of experiments, our findings show that the parameters that characterize the adversary model significantly affect the evaluation of the generalization algorithms. This implies that when a generalization algorithm is evaluated it is necessary to estimate realistic values for these parameters. Indeed, an error in the estimation may lead to misleading results.

5.4 Impact of the Deployment Model on the Evaluation of the Generalization Algorithms

We now show a set of experimental results designed to evaluate the impact of the deployment model on the evaluation of the generalization algorithms.

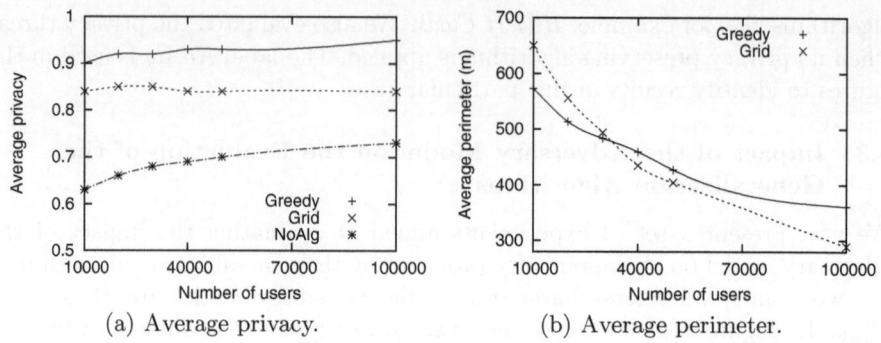

(a) Average privacy.   (b) Average perimeter.

Fig. 4. Performance evaluation for different values of the total population

Figure 4(a) shows that the average privacy obtained with *Greedy* and *Grid* is not significantly affected by the size of the total population. Indeed, both algorithms, independently from the total number of users, try to have generalized locations that cover the location of k users, so the privacy of the requests is not affected. However, when the density is high, the two algorithms can generalize to a small area, while when the density is low, a larger area is necessary to cover the location of k users (see Figure 4(b)). On the contrary, the privacy obtained when no generalization is performed is significantly affected by the total population. Indeed, a higher density increases the probability of different users to be in the same location and hence it increases privacy also if the requests are not generalized.

The set of tests reported in in Figure 5 compares the privacy achieved by the *Greedy* algorithm on the two datasets for different values of k and for different values of QoS. The experiments on MRM were repeated trying also larger values for the QoS threshold ($maxP = 2000$ and $maxP = 4000$), so three different versions of MRM appear in the figures. In order to focus on these parameters only, in these tests the probability of identification was set to the same value for any place ($P_{id-in} = P_{id-out} = 0.1$), and for the MRM dataset the issuer of the requests was randomly chosen only among those that stay in the simulation for 3 hours, ignoring the ones staying for much shorter time that inevitably are part of this dataset. This setting allowed us to compare the results on the two datasets using the same average length of traces identified by the adversary.

Figure 5(a) shows that the average privacy of the algorithm evaluated on the AB dataset is much higher than on the MRM dataset. This is mainly motivated by the fact that in AB users tend to concentrate in a few locations (the entertainment places) and this enhances privacy. This is also confirmed by a similar test performed without using any generalization of locations; we obtained values constantly higher for the AB dataset (the average privacy is 0.67 in AB and 0.55 in MRM).

In Figure 5(b) we show the QoS achieved by the algorithm in the two datasets with respect to the average privacy achieved. This result confirms that the level of privacy evaluated on the AB dataset using small values of k and $maxP$ for

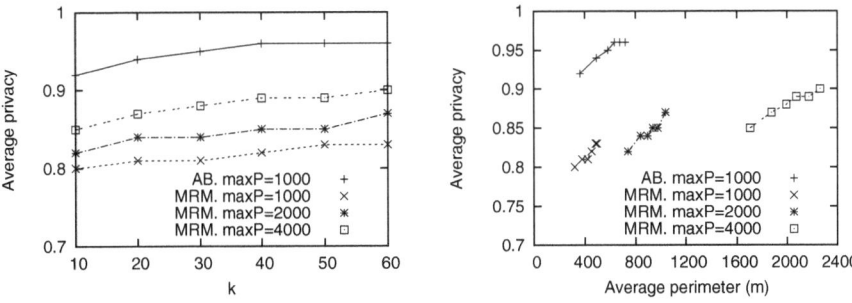

(a) Average privacy as a function of the level of indistinguishability k.

(b) Average privacy as a function of the average perimeter.

Fig. 5. Evaluation of the *Greedy* algorithm using AB and MRM data sets. $P_{id-in} = P_{id-out} = 0.1$.

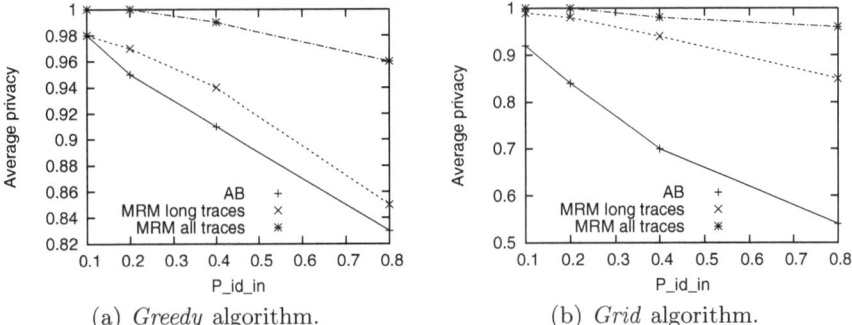

(a) *Greedy* algorithm.

(b) *Grid* algorithm.

Fig. 6. Average privacy using AB and MRM data sets. $P_{id-out} = P_{id-in}/10$.

the algorithm cannot be observed on the MRM dataset even with much higher values for these parameters.

From the experiments shown in Figure 5 we can conclude that if the MRM dataset is used as a benchmark to estimate the values of k and $maxP$ that are necessary to provide a desired average level of privacy, then the results will suggest the use of values that are over-protective. As a consequence, it is possible that the service will exhibit a much lower QoS than the one that could be achieved with the same algorithm.

The above results may still support the safety of using MRM, since according to what we have seen above a technique achieving a certain level of privacy may only do better in a real scenario. However, our second set of experiments shows that this is not the case.

In Figure 6 we show the results we obtained by varying the probability of identification. For this test, we considered two sets of issuers in the MRM data set. One set is composed by users that stay in the simulation for 3 hours, (MRM *long traces*, in Figure 6), while the other contains issuers randomly chosen in the entire set of users (*MRM all traces*, in Figure 6), hence including users staying in the simulation for a much shorter time.

In Figure 6(a) and 6(b) we can observe that the execution on the *MRM* dataset leads to evaluate a privacy level that is higher than the one obtained on the *AB* dataset. In particular, the evaluation of the *Grid* algorithm using the *MRM* dataset (Figure 6(b)), would suggest that the algorithm is able to provide a high privacy protection. However, when evaluating the same algorithm using the more realistic dataset *AB*, this conclusion seems to be incorrect. In this case, the evaluation on the *MRM* dataset may lead to underestimate the privacy risk, and hence to deploy services based on generalization algorithms that may not provide the minimum required level of privacy.

6   Open Problems

As seen from the previous sections, progress has been made in protecting users' privacy in using location based services. However, much research is still needed. In this section we discuss some open problems that are immediately related to the anonymity-based techniques we discussed in this contribution.

Recognizing the dynamic role of quasi-identifiers and of private information. All the techniques proposed so far in the literature assume that the informations in the request acting as quasi-identifier or as private information do not change among different requests. However, it should be observed that this may not always be the case. Indeed, in a realistic scenario, only some locations can act as quasi-identifiers, and, similarly, only some service requests contain private information (location and/or service parameters). The proper recognition of the role of information in the requests is crucial in designing an effective defense technique. Indeed, over conservative assumptions lead to quality of service degradation, and ignoring the role of data as quasi-identifier or private information in a request leads to privacy violation.

Pattern-based quasi-identifiers. In the historical case the adversary can observe some movement patterns. In this case, even if a single request contains no information acting as quasi-identifier, the sequence of movements can lead to the identification of the issuer. Consider the following example: a user issues several linkable requests from her home (location A) and workplace (location B). Assume that, since the requests are generalized to areas containing public places, the adversary cannot restrict the set of possible issuers by considering each single request. However, if the adversary is able to extract a movement pattern from the requests, he can infer that the issuer most probably lives in location A and works in location B, and this information is a quasi-identifier [12,36].

Personalization of the degree of anonymity. In our discussion we never considered issues related to the personalization of defense parameters, as for example, the degree of anonymity k to be enforced. Some approaches (e.g. [9]) actually explicitly allow different users to specify different values of k. A natural question is if the other techniques can be applied and can be considered safe even in this case. Once again, to answer this question it is essential to consider which knowledge an adversary may obtain. The degree of anonymity k desired by each user at the time of a request is not assumed to be known by the adversary (even in the def-aware attacks) in the presented algorithms, hence the algorithms that are safe against the corresponding attacks remain safe even when personalized values for k are considered.

However, it may be reasonable to consider attacks in which the adversary may obtain information about k. In the multiple-issuer case, the adversary may use, for example, data mining techniques to figure out the k value. Example 8 shows that, in such a scenario, the presented algorithms need to be extended in order to provide an effective defense.

Example 8. Alice issues a request r asking a degree of anonymity $k = 2$. Using a defense algorithm against def-aware attacks, r is generalized to the request r' that has a spatiotemporal region containing only Alice and Bob. Since the generalization algorithm is safe against def-aware attacks, if r were issued by Bob with $k = 2$, then it would be generalized to r'. However, if the adversary knows that Bob always issues requests with $k \geq 3$, then he knows that if the issuer of r were Bob, the request would have been generalized to a request r'' different from r', because the spatiotemporal region of r'' should include at least 3 users. Hence the adversary would identify Alice as the issuer of r'.

Deployment-aware data generator. Earlier, we claimed that the experimental evaluation of LBS privacy preserving techniques should be based on user movement datasets obtained through simulations *tailored* to the specific deployment scenario of the target services. Our results support our thesis for the class of LBS known as friend-finder services, for defense techniques based on spatial cloaking, and for attack models that include the possibility for the adversary to occasionally recognize people in certain locations. These results can be extended to other types of LBS, other defense techniques, and various types of attacks. Thus, we believe a significant effort should be devoted to the development of new flexible and efficient context-aware user movement simulators, as well as to the collection of real data, possibly even in an aggregated form, to properly tune the simulations. In our opinion this is a necessary step to have significant common benchmarks to evaluate LBS privacy preserving techniques.

7   Conclusion

In this contribution, we introduced the privacy problem in LBS by categorizing both attacks and existing defense techniques. We then discussed the use of anonymity for protection, focusing on the notion of historical k-anonymity and on the techniques to ensure this form of anonymity. Finally, we provided a performance evaluation of these techniques depending on the adversary model and on the specific service deployment model. Based on our extensive work on the simulation environment, we believe that the design of realistic simulations for specific services, possibly driven by real data, is today one of the main challenges in this field, since proposed defenses need serious evaluation, and theoretical validation is important but has several limits, mainly due to the conservative assumptions that seem very hard to avoid.

Acknowledgments

This work was partially supported by National Science Foundation (NSF) under grant N. CNS-0716567, and by Italian MIUR under grants InterLink-II04C0EC1D and PRIN-2007F9437X.

References

1. Barkhuus, L., Dey, A.: Location-based services for mobile telephony: a study of users privacy concerns. In: Proc. of the 9th International Conference on Human-Computer Interaction, pp. 709–712. IOS Press, Amsterdam (2003)
2. Ciriani, V., di Vimercati, S.D.C., Foresti, S., Samarati, P.: k-Anonymity. In: Secure Data Management in Decentralized Systems. Springer, Heidelberg (2007)
3. Bettini, C., Wang, X.S., Jajodia, S.: How anonymous is k-anonymous? look at your quasi-id. In: Jonker, W., Petković, M. (eds.) SDM 2008. LNCS, vol. 5159, pp. 1–15. Springer, Heidelberg (2008)
4. Dalenius, T.: Finding a needle in a haystack - or identifying anonymous census record. Journal of Official Statistics 2(3), 329–336 (1986)
5. Abul, O., Bonchi, F., Nanni, M.: Never walk alone: Uncertainty for anonymity in moving objects databases. In: Proc. of the 24th International Conference on Data Engineering, pp. 376–386. IEEE Computer Society, Los Alamitos (2008)
6. Bettini, C., Mascetti, S., Wang, X.S., Jajodia, S.: Anonymity in location-based services: towards a general framework. In: Proc. of the 8th International Conference on Mobile Data Management, pp. 69–76. IEEE Computer Society, Los Alamitos (2007)
7. Gruteser, M., Grunwald, D.: Anonymous usage of location-based services through spatial and temporal cloaking. In: Proc. of the 1st International Conference on Mobile Systems, Applications and Services, pp. 31–42. The USENIX Association (2003)
8. Kalnis, P., Ghinita, G., Mouratidis, K., Papadias, D.: Preventing location-based identity inference in anonymous spatial queries. IEEE Transactions on Knowledge and Data Engineering 19(12), 1719–1733 (2007)
9. Mokbel, M.F., Chow, C.Y., Aref, W.G.: The new casper: query processing for location services without compromising privacy. In: Proc. of the 32nd International Conference on Very Large Data Bases, VLDB Endowment, pp. 763–774 (2006)
10. Beresford, A.R., Stajano, F.: Mix zones: User privacy in location-aware services. In: Proc. of the 2nd Annual Conference on Pervasive Computing and Communications, pp. 127–131. IEEE Computer Society, Los Alamitos (2004)
11. Hoh, B., Gruteser, M.: Protecting location privacy through path confusion. In: Proc. of the First International Conference on Security and Privacy for Emerging Areas in Communications Networks, pp. 194–205. IEEE Computer Society, Los Alamitos (2005)
12. Bettini, C., Wang, X.S., Jajodia, S.: Protecting privacy against location-based personal identification. In: Jonker, W., Petković, M. (eds.) SDM 2005. LNCS, vol. 3674, pp. 185–199. Springer, Heidelberg (2005)

13. Machanavajjhala, A., Gehrke, J., Kifer, D., Venkitasubramaniam, M.: l-Diversity: Privacy Beyond k-Anonymity. In: Proceedings of the 22nd International Conference on Data Engineering, p. 24. IEEE Computer Society, Los Alamitos (2006)
14. Mascetti, S., Bettini, C., Freni, D., Wang, X.S.: Spatial generalization algorithms for LBS privacy preservation. Journal of Location Based Services 2(1), 179–207 (2008)
15. Gedik, B., Liu, L.: Protecting location privacy with personalized k-anonymity: Architecture and algorithms. IEEE Transactions on Mobile Computing 7(1), 1–18 (2008)
16. Chow, C.Y., Mokbel, M.F., Liu, X.: A peer-to-peer spatial cloaking algorithm for anonymous location-based service. In: Proc. of the 14th International Symposium on Geographic Information Systems, pp. 171–178. ACM, New York (2006)
17. Ghinita, G., Kalnis, P., Skiadopoulos, S.: Prive: anonymous location-based queries in distributed mobile systems. In: Proc. of the 16th international conference on World Wide Web, pp. 371–380. ACM Press, New York (2007)
18. Ghinita, G., Kalnis, P., Skiadopoulos, S.: Mobihide: A mobile peer-to-peer system for anonymous location-based queries. In: Papadias, D., Zhang, D., Kollios, G. (eds.) SSTD 2007. LNCS, vol. 4605, pp. 221–238. Springer, Heidelberg (2007)
19. Hu, H., Xu, J.: Non-exposure location anonymity. In: Proc. of the 25th International Conference on Data Engineering, pp. 1120–1131. IEEE Computer Society, Los Alamitos (2009)
20. Chow, C.Y., Mokbel, M.: Enabling private continuous queries for revealed user locations. In: Papadias, D., Zhang, D., Kollios, G. (eds.) SSTD 2007. LNCS, vol. 4605, pp. 258–275. Springer, Heidelberg (2007)
21. Xu, T., Cai, Y.: Location anonymity in continuous location-based services. In: Proc. of ACM International Symposium on Advances in Geographic Information Systems, p. 39. ACM Press, New York (2007)
22. Mascetti, S., Bettini, C., Wang, X.S., Freni, D., Jajodia, S.: ProvidentHider: an algorithm to preserve historical k-anonymity in lbs. In: Proc. of the 10th International Conference on Mobile Data Management, pp. 172–181. IEEE Computer Society, Los Alamitos (2009)
23. Bettini, C., Jajodia, S., Pareschi, L.: Anonymity and diversity in LBS: a preliminary investigation. In: Proc. of the 5th International Conference on Pervasive Computing and Communications, pp. 577–580. IEEE Computer Society, Los Alamitos (2007)
24. Riboni, D., Pareschi, L., Bettini, C., Jajodia, S.: Preserving anonymity of recurrent location-based queries. In: Proc. of 16th International Symposium on Temporal Representation and Reasoning. IEEE Computer Society, Los Alamitos (2009)
25. Gruteser, M., Liu, X.: Protecting privacy in continuous location-tracking applications. IEEE Security & Privacy 2(2), 28–34 (2004)
26. Duckham, M., Kulik, L.: A formal model of obfuscation and negotiation for location privacy. In: Gellersen, H.-W., Want, R., Schmidt, A. (eds.) PERVASIVE 2005. LNCS, vol. 3468, pp. 152–170. Springer, Heidelberg (2005)
27. Kido, H., Yanagisawa, Y., Satoh, T.: Protection of location privacy using dummies for location-based services. In: Proc. of the 21st International Conference on Data Engineering Workshops, p. 1248. IEEE Computer Society, Los Alamitos (2005)
28. Ardagna, C.A., Cremonini, M., Damiani, E., di Vimercati, S.D.C., Samarati, P.: Location privacy protection through obfuscation-based techniques. In: Barker, S., Ahn, G.-J. (eds.) Data and Applications Security 2007. LNCS, vol. 4602, pp. 47–60. Springer, Heidelberg (2007)

29. Yiu, M.L., Jensen, C.S., Huang, X., Lu, H.: Spacetwist: Managing the trade-offs among location privacy, query performance, and query accuracy in mobile services. In: Proc. of the 24th International Conference on Data Engineering, pp. 366–375. IEEE Computer Society, Los Alamitos (2008)
30. Ghinita, G., Kalnis, P., Khoshgozaran, A., Shahabi, C., Tan, K.L.: Private queries in location based services: Anonymizers are not necessary. In: Proc. of SIGMOD, pp. 121–132. ACM Press, New York (2008)
31. Mascetti, S., Bettini, C., Freni, D., Wang, X.S., Jajodia, S.: Privacy-aware proximity based services. In: Proc. of the 10th International Conference on Mobile Data Management, pp. 31–40. IEEE Computer Society, Los Alamitos (2009)
32. Brinkhoff, T.: A framework for generating network-based moving objects. GeoInformatica 6(2), 153–180 (2002)
33. Martin, M., Nurmi, P.: A generic large scale simulator for ubiquitous computing. In: Proc. of the 3rd Conference on Mobile and Ubiquitous Systems: Networks and Services. IEEE Computer Society, Los Alamitos (2006)
34. Mascetti, S., Freni, D., Bettini, C., Wang, X.S., Jajodia, S.: On the impact of user movement simulations in the evaluation of LBS privacy-preserving techniques. In: Proc. of the International Workshop on Privacy in Location-Based Applications, Malaga, Spain. CEUR-WS, vol. 397, pp. 61–80 (2008)
35. Vyahhi, N., Bakiras, S., Kalnis, P., Ghinita, G.: Tracking moving objects in anonymized trajectories. In: Bhowmick, S.S., Küng, J., Wagner, R. (eds.) DEXA 2008. LNCS, vol. 5181, pp. 158–171. Springer, Heidelberg (2008)
36. Golle, P., Partridge, K.: On the anonymity of home/work location pairs. In: Tokuda, H., Beigl, M., Friday, A., Bernheim Brush, A.J., Tobe, Y. (eds.) Pervasive 2009. LNCS, vol. 5538, pp. 390–397. Springer, Heidelberg (2009)

Location Privacy Techniques in Client-Server Architectures

Christian S. Jensen[1,2], Hua Lu[2], and Man Lung Yiu[2]

[1] Google Inc., Mountain View, CA 94043, USA
[2] Department of Computer Science, Aalborg University, Denmark
{csj,luhua,mly}@cs.aau.dk

Abstract. A typical location-based service returns nearby points of interest in response to a user location. As such services are becoming increasingly available and popular, location privacy emerges as an important issue. In a system that does not offer location privacy, users must disclose their exact locations in order to receive the desired services. We view location privacy as an enabling technology that may lead to increased use of location-based services.

In this chapter, we consider location privacy techniques that work in traditional client-server architectures without any trusted components other than the client's mobile device. Such techniques have important advantages. First, they are relatively easy to implement because they do not rely on any trusted third-party components. Second, they have potential for wide application, as the client-server architecture remains dominant for web services. Third, their effectiveness is independent of the distribution of other users, unlike the k-anonymity approach.

The chapter characterizes the privacy models assumed by existing techniques and categorizes these according to their approach. The techniques are then covered in turn according to their category. The first category of techniques enlarge the client's position into a region before it is sent to the server. Next, dummy-based techniques hide the user's true location among fake locations, called dummies. In progressive retrieval, candidate results are retrieved iteratively from the server, without disclosing the exact user location. Finally, transformation-based techniques employ cryptographic transformations so that the service provider is unable to decipher the exact user locations. We end by pointing out promising directions and open problems.

1   Introduction

The Internet is rapidly becoming mobile. An infrastructure is emerging that encompasses large numbers of users equipped with mobile terminals that posses geo-positioning capabilities (e.g., built-in GPS receivers) and data communication capabilities. Thus, location-based services (LBS) are increasingly becoming available. These return results relative to the users' locations. An example service returns the gas station nearest to the location of a user. Another example is a service that returns all restaurants within 2 km of the user's location.

C. Bettini et al. (Eds.): Privacy in Location-Based Applications, LNCS 5599, pp. 31–58, 2009.

To receive such services, the users must disclose their locations to the service provider. Users may be uncomfortable disclosing their exact locations to an untrusted service provider that may misuse the knowledge of the users' locations [1]. We view location privacy as an enabling technology for the diffusion of the mobile Internet and the proliferation of location-based services. By offering users the ability to choose different levels of location privacy, users are encouraged to use mobile services more often.

Some existing location privacy solutions assume the presence of a *centralized* third-party anonymizer that is aware of all users' locations. This trusted anonymizer serves as an intermediary in-between the users and the service provider. However, such an anonymizer may not always be practical, and it may itself present security, performance, and privacy problems. For example, the anonymizer represents a single-point-of-attack for hackers. Also, the anonymizer is prone to becoming a performance bottleneck because it may need to serve a large number of users.

In contrast, the techniques covered in this chapter assume a client-server architecture without any third-party anonymizer. We therefore call these *decentralized* solutions. The decentralized solutions are motivated by several considerations. First, the client-server architecture is widely used by today's location-based services. This popularity affords decentralized solutions wide applicability.

Second, a mobile terminal in a decentralized solution does not need to keep an anonymizer up to date with its location at all times; the terminal only issues queries to the server on demand. The anonymizer of a centralized solution needs to maintain up-to-date locations of all mobile terminals in order to perform cloaking for the small fraction of users that are issuing queries at any point in time.

Third, the setting of this chapter is based on the seemingly realistic assumptions that an adversary knows what the service provider knows, i.e., the identity of the user who issues a query and the parameters and result of the query. Specifically, we assume that users must register with the service provider to receive services; and we assume that users are not required to report their latest locations continuously.

In the next section, we provide an overview of decentralized solutions found in the literature.

2   Overview of Client-Server Solutions

The privacy models of existing solutions can be broadly classified into two types: identity privacy and location privacy.

The *identity privacy* model [2] assumes that (i) an untrusted party has access to a location database that records the exact location of each user in the population of users and (ii) that service users are anonymous. If a service user discloses her exact location to the untrusted party, that party may be able to retrieve the user's identity from the location database. In this setting, which this chapter does not consider, the location of a user is obfuscated in order to preserve the anonymity of the user.

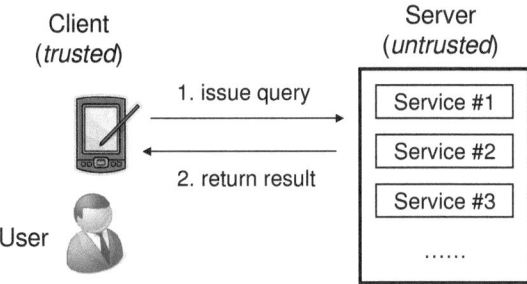

Fig. 1. Client-Server Architecture

This chapter is devoted to the *location privacy* model, which assumes that untrusted parties know the user's identity, but not the user's location. This model fits well with services where a user must log in before using the services. Examples include location-based services available in Google's Android Market[1]. Also, FireEagle[2] by Yahoo! enables users to share their locations with their friends, allowing them to specify the preciseness of the shared locations (e.g., exact location, city of the location, or undisclosed location).

Under the above model, we study privacy solutions that simply assume a *client-server* architecture and that apply to *snapshot* queries based on *the user's location*. In other words, we consider neither the privacy of continuous queries nor of a user's trajectory. Figure 1 illustrates the client-server architecture, in which the client is trusted, but the server (including its services) is not trusted. It does not rely on peer-to-peer communication among the clients, and nor does it employ a trusted third-party anonymizer.

Existing solutions for the location privacy model can be classified into four categories.

- *Query enlargement* techniques [3, 4, 5, 6, 7] (Section 3) enlarge the client's exact position into a region before sending it to the server.
- *Dummy-based* techniques [8, 9] (Section 4) generate dummies (i.e., fake locations) at the client and then send them together with the exact user location to the service provider, thus hiding the user location among the dummies.
- *Progressive retrieval* techniques [10, 11, 12] (Section 5) iteratively retrieve candidate results from the server, without disclosing the exact user location.
- *Transformation-based* techniques [13, 14] (Section 6) employ cryptographic transformation so that the service provider is unable to decipher the exact user locations, while providing the clients with decryption functionality so that they can derive the actual results.

Table 1 offers a summary of specific location privacy solutions that belong to the above categories. Six features are covered: (i) the nature of the domain space, (ii) the privacy measure, (iii) the types of queries supported, (iv) whether

---

[1] http://www.android.com
[2] http://fireeagle.yahoo.net

Table 1. Features of Various Location Privacy Techniques

Method	Domain Space	Privacy Measure	Supported Queries	Exact Result	Accuracy Guarantee	Impl. Difficulty
[3]	Euclidean	Area-based	Range	Yes	Yes	Medium
[4]	Euclidean	Area-based	Range, kNN	Yes	Yes	Medium
[5, 6]	Euclidean	Area-based	Range, kNN	Yes	Yes	Medium
[7]	Euclidean	Area-based	Proximity	No	No	Medium
[8]	Euclidean	Size-based	Range, kNN	Yes	Yes	Low
[9]	Euclidean	Size and Area	Range, kNN	Yes	Yes	Low
[10, 11]	Network	Size-based	1NN	Yes/No	Yes/No	Medium
[12]	Euclidean	Distance-based	kNN	Yes	Yes	Low
[12]g	Euclidean	Distance-based	kNN	No	Yes	Low
[13]	Euclidean	Full-domain	kNN	No	No	Medium
[14]	Euclidean	Full-domain	1NN	Yes	Yes	High

exact results can be retrieved, (v) whether result accuracy guarantees are given (for approximate results), and (vi) the difficulty of implementing the solution.

The domain space used by Duckham and Kulik [10, 11] is modeled by a graph that represents a road network. All the other work focus on the Euclidean space. No existing solution is applicable to both Euclidean space and network space simultaneously.

The privacy measure, i.e., the means of quantifying the privacy afforded a user, of the solutions can be classified into four categories. First, in the *area-based* measures [3, 4, 5, 6, 7], the privacy of the user is measured by the area (or a derivative of it) of the region that contains the user's location. Second, the *size-based* measures [8, 10, 11] simply express the privacy as the cardinality of a discrete set of locations that contains the user's location. The work of Lu et al. [9] employs a hybrid that builds on the size-based and area-based measures. Third, the *distance-based* privacy measures [12] capture the expected distance of the user's location from the adversary's estimate. Fourth, the *full-domain* privacy measures [13, 14] ensure that the adversary cannot learn any information on the user's location, as it is transformed into another space.

An interesting issue is to examine whether a particular privacy model is applicable to other solutions. Among the solutions covered, the full-domain measure is applicable only to the solutions in references [13, 14]. The distance-based measure is applicable to the solutions in references [3, 4, 5, 6, 8, 9, 10, 11]. It can also be noted that the area-based measures cannot be applied to the solutions in references [8, 10, 11] that use a discrete set of points, whereas the size-based measure is inapplicable to the solutions in references [3, 4, 5, 6] that use a single continuous region for cloaking.

The typical queries that underlie location-based services are the range query and the k-nearest neighbor query. Given a dataset P (of points of interest, or data points) and a query region W, the *range query* retrieves each object $o \in P$ such that o intersects with W. Given a set P and a query point q, the *k-nearest neighbor query* retrieves k objects from P such that their distances from q are minimized. it follows from Table 1 that some solutions support range queries only, some support k-nearest neighbor queries only, and some support both. It is worth noticing that the methods in references [10, 11, 14] support only the nearest neighbor query (i.e., the special case with $k = 1$). In addition, proximity based queries (e.g., finding those of my friends that are close to me) are supported [7].

We cover two aspects that relate to the quality of a query result: whether it either is or contains the exact result, and, if not, whether an accuracy guarantee is provided. We observe that most of the existing solutions guarantee that their results are supersets of the actual results, thus allowing the client to obtain the exact result. The solutions of Duckham and Kulik [10, 11] ensure that the exact result is returned only if the user agrees to reveal a sufficiently accurate obfuscation of her location. The table uses "Yes/No" to capture this conditional property. Otherwise, the solution does not guarantee the accuracy of the returned result (thus the corresponding "Yes/No"). Yiu et al. [12] propose a solution that offers exact results and thus accuracy guarantees. In addition, an extension that utilizes so-called granular search for improving performance returns approximate results with user-controlled accuracy guarantees. In the table, this extension is called [12]g. The work of Khoshgozaran and Shahabi [13] does not provide result accuracy guarantees, and it cannot support exact result retrieval.

The aspect concerns the difficulty of implementing and deploying the proposed solutions. The solutions in references [8, 9, 12] are easy to implement as they reuse existing location-based operations that can be assumed to be available in location based servers. The solutions in references [3, 4, 5, 6, 7, 10, 11] have medium implementation difficulty as they apply specialized geometric search algorithms. The method of Khoshgozaran and Shahabi [13] also has medium implementation difficulty because a Hilbert curve transformation function needs to be used by the client. The solution of Ghinita et al. [14] has high implementation difficulty as both the client and the server have to run a protocol for private information retrieval.

3   Query Enlargement Techniques

A straightforward way of protecting an exact user location in a service request is to replace the user location by a region that contains the location. We call the solutions that adopt this tack *Query Enlargement Techniques*. Unlike centralized cloaking solutions, the query enlargement techniques considered here do not require any trusted third-party component.

3.1   Cloaking Agent-Based Technique

Cheng et al. [3] assume a setting in which the data points are not the typical, static points of interest such as restaurants, but are the locations of other users. thus, user requests are intended to retrieve private data rather than public data, as do all other techniques covered in this chapter.

In this setting, the service quality may degrade when the spatial and temporal information sent to the service provider is at a coarse granularity. Motivated by this, Cheng et al. [3] proposed a framework for balancing the user location privacy and quality of service requested.

Architecture

The proposed architecture is illustrated in Figure 2. It encompasses of a crucial component, the *cloaking agent*. The cloaking agent is not necessarily a third-party component—it can also be implemented directly on the client side, i.e., on the user's device. For this reason, we cover this technique.

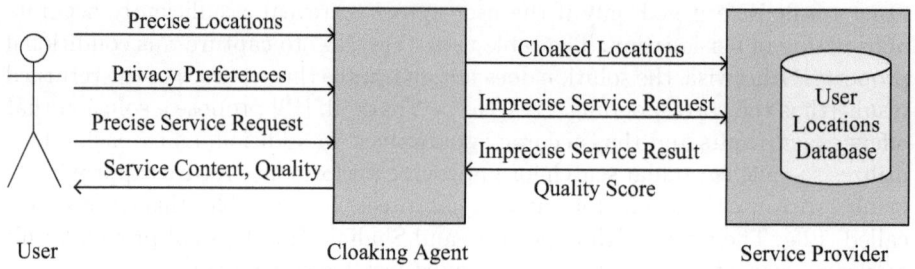

Fig. 2. Cloaking Agent-Based Architecture for Privacy and Service Quality Tradeoff

In particular, the cloaking agent receives precise locations and privacy preferences from a user, introduces uncertainty into the user's locations according to the privacy preferences, and reports the uncertain locations to the database at the service provider side.

When the user issues a service request with an exact location, the request is passed to the cloaking agent where it is translated into an imprecise service request with a cloaked location obtained according to the user's privacy preferences as known by the agent.

The imprecise service request is then sent to the service provider where it is processed using the uncertain user locations stored in its database, yielding an imprecise service result. The imprecise result, together with a score quantifying the service quality, is then sent back to the cloaking agent. The cloaking agent delivers the service result and the quality measurement to the user, who is allowed to adjust the privacy preferences based on the service and quality received.

Privacy Model

Cheng et al. [3] base the specification of location privacy preferences on a probabilistic location cloaking model. Assume that n users, namely $S_1, S_2, \ldots, S_n$, are registered in the system. Let $L_i(t)$ be the exact location of user S_i at time t. Instead of reporting $L_i(t)$, the user S_i reports a closed uncertainty region $U_i(t)$ to the service provider, such that $L_i(t)$ has a uniform probability distribution in $U_i(t)$.

A user is allowed to control the degree of location privacy in two ways. First, a user can specify the desired area of the uncertainty region, i.e., $Area(U_i(t))$. In general, the larger the value of $Area(U_i(t))$, the higher the location privacy. The idea is that it is more difficult for the adversary to determine the user's exact location $L_i(t)$ the larger the uncertainty region $U_i(t)$ becomes.

Second, a user can specify the desired coverage of sensitive regions. When the user is in a sensitive region, e.g., at a psychology clinic, she does not want to release the location information. However, if the user's uncertainty region happens to overlap with the clinic by a high percentage (e.g., 90%), it becomes easy for the adversary to guess that the user is at the clinic. To overcome this problem, a user may specify a coverage value based on the equation below that is not to be exceeded.

$$Coverage = \frac{Area(sensitive\ regions \cap U_i(t))}{Area(U_i(t))}$$

Query Processing

Cheng et al. focus on the processing of range queries. Based on their probabilistic location cloaking model, a range query from user S_i is translated by the cloaking agent into an imprecise location-based range query (ILRQ). An ILRQ issued at time t returns the set $\{(S_j, p_j) \mid j \neq i \wedge j \in [1, n]\}$, where $p_j > 0$ is the probability that S_j is located within $U_i(t)$ at time t. As mentioned, such a query concerns other users' locations, not public points of interest such as restaurants.

An ILRQ is processed by the service provider in three phases: (i) The pruning phase eliminates objects whose uncertainty regions do not overlap with the ILRQ. (ii) The transformation phase transforms an ILRQ into subqueries. For each possible location $(u, v) \in U_i(t)$ of S_i, a subquery is generated to find those unpruned objects whose uncertainty regions overlap with the circle centered at (u, v) and with radius r (specified in the original query), denoted as $C((u, v), r)$. (iii) The evaluation phase evaluates each subquery, by computing the actual probability that each remaining object satisfies the ILRQ. The probability of object S_j satisfying a subquery located at (u, v) is given as:

$$p_j(u, v) = \frac{Area(U_j(t) \cap C((u, v), r))}{Area(U_j(t))}$$

The results of all subqueries are combined as the answer to the original ILRQ.

From the location privacy preference specification above, it is easy to see that better user location privacy results from using a larger uncertainty region. However, simply increasing the uncertainty region inevitably hurts the service quality. Specifically, the use of larges uncertainty regions tends to retrieve more objects with lower probabilities. To enable trade-offs between privacy and service quality, a service quality metric is proposed.

Result Quality

Assume that an ILRQ from user S is partitioned into B subqueries that correspond to B locations among A_1 to A_B. Let the probability that S is located at A_k $(1 \leq k \leq B)$ be $p_k(S)$. The result of the subquery at A_k is R_k, while $R = \bigcup_{k=1}^{B} R_k$. The quality score of the ILRQ is defined as follows:

$$Query\ score = \sum_{k=1}^{B} p_k(S) \cdot \frac{|R_k|}{|R|}$$

The score varies between 0, the lowest quality, and 1, the highest quality. When a user S receives a query result and its corresponding score, she can adjust privacy preferences stored in the cloaking agent according to her expectation and the score value.

Benefits and Limitations

The proposed solution has two advantages. First, it allows flexibility on the architecture, as the cloaking agent can be part of the client or can be a separate third party. Second, it offers quantifies location privacy and service quality, which allows users to make trade-off according to their needs.

Nevertheless, the cloaking agent based solution also suffer from some disadvantages. First, it is assumed that the service provider knows all possible locations where a user can be. This is exploited in the query transformation and query quality score calculation. If there are many such locations, the query transformation, the query evaluation, and the quality score calculation can all be very expensive. And if there are few such locations, the location privacy is not well protected. Second, it may be difficult for a user to understand well the exact meaning of service quality scores, which therefore may reduce the utility of such scores.

3.2   Spatial Obfuscation Techniques

Ardagna et al. [4] propose a straightforward and intuitive way to express user location privacy preferences using obfuscated circles. Due to assumed measurement accuracy limitations, a user location is represented as a circular region $C((x_c, y_c), r_{meas})$ (i.e., centered at (x_c, y_c) and with radius r_{meas}). The possible user locations are assumed to be uniformly distributed within that region.

Measurement of Privacy and Accuracy

To support multiple location obfuscation techniques, an attribute λ is first introduced to represent a *relative privacy preference*, which is derived according to the following formula:

$$\lambda = \frac{max(r_{meas}, r_{min})^2}{r_{meas}^2} - 1$$

Here, r_{meas} represents measurement accuracy, i.e., the radius of a measured circular region modeling the user location; r_{min} is the minimum distance specified by a user to express her privacy preference [11]. For example, "1 mile" indicates that the user requires her location to be represented by a circular region with a radius of at least 1 mile. The term $max(r_{meas}, r_{min})$ is used because it is possible that $r_{min} < r_{meas}$ because the measurement accuracy may be unknown to the user.

When $r_{meas} \geq r_{min}$, $\lambda = 0$, indicating that the user privacy preference is already satisfied as the measured radius exceeds the preferred minimum distance. In this case, no location obfuscation is needed. When $r_{meas} < r_{min}$, $\lambda > 0$, reflecting the degree to which the location accuracy is to be degraded to protect the user according to the user's privacy preference. In this case, obfuscation is needed.

To measure the accuracy of an obfuscated region, a technology-independent metric, called *relevance*, is defined as a value $\mathcal{R} \in (0, 1]$. The relevance is 1 when the user location has the best accuracy, and its value is close to 0 when the user location is considered too inaccurate to be used by the service provider. The value $1 - \mathcal{R}$ is accordingly the *location privacy* offered by an obfuscated location.

The privacy management solution of Ardagna et al. embodies two crucial relevance values. The *initial relevance* ($\mathcal{R}_{Init}$) is the measure of the accuracy of a user location as obtained using some positioning technology. The *final relevance* ($\mathcal{R}_{Final}$) is the measure of the accuracy of the final obfuscated region that satisfies a relative privacy preference λ. let r_{opt} be the measurement radius corresponding to the best accuracy of a positioning technology. The initial and final relevance are calculated as follows:

$$\mathcal{R}_{Init} = \frac{r_{opt}^2}{r_{meas}^2} \qquad \mathcal{R}_{Final} = \frac{\mathcal{R}_{Init}}{\lambda + 1}$$

Obfuscation Operators

To derive $\mathcal{R}_{Final}$ from $\mathcal{R}_{Init}$, three basic obfuscation operators are defined on circular regions. First, the *Enlarge* operator (E) enlarges the radius of a region. Second, the *Shift* operator (S) shifts the center of a region. Third, the Reduce operator (R) reduces the radius of a region.

An example of E obfuscation operator is illustrated in Figure 3(a). Here the initial radius r is increased to $r' > r$. Let $\mathcal{R}$ and $\mathcal{R}'$ be the relevances before and after the operator, respectively. Then $\mathcal{R}'$ is derived from $\mathcal{R}$ as follows:

$$\mathcal{R}' = \frac{f_{r'}(x, y)}{f_r(x, y)} \cdot \mathcal{R} = \frac{r^2}{r'^2} \cdot \mathcal{R}$$

In the formula, $f_r(x, y)$ ($f_{r'}(x, y)$) is the joint probability density function (pdf) of an exact user location to be in the circular region indicated by r (r'). Note that $\mathcal{R}' < \mathcal{R}$ as $r' > r$. Therefore, $1 - \mathcal{R}' > 1 - \mathcal{R}$, which means that the location privacy is increased by the Enlarge operator.

An example of the S obfuscation operator is illustrated in Figure 3(b). Here the initial center is shifted by a vector ($\Delta x, \Delta y$) of length d. The relevance of the result of applying the operator is derived as follows:

$$\mathcal{R}' = P((x_u, y_u) \in C_{Init} \cap C_{Final}) \cdot P((x, y) \in C_{Init} \cap C_{Final})$$
$$= \frac{Area(C_{Init} \cap C_{Final})^2}{Area((x_c, y_c), r)^2} \cdot \mathcal{R}$$

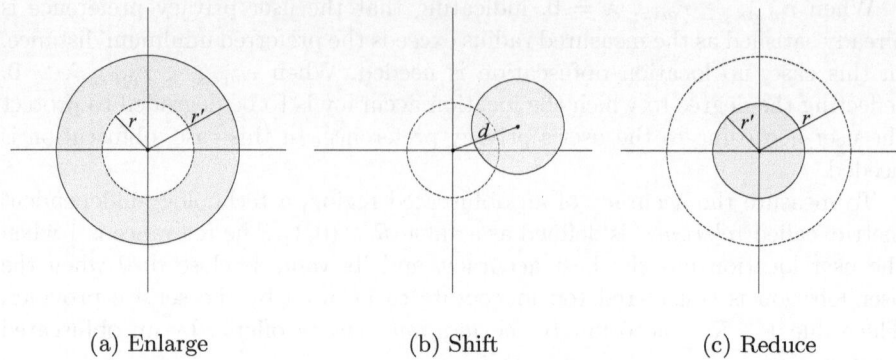

(a) Enlarge   (b) Shift   (c) Reduce

Fig. 3. Basic Obfuscation Operators

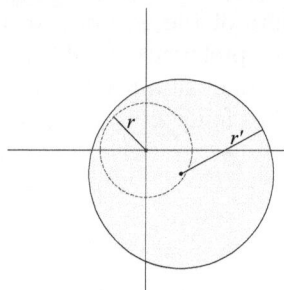

Fig. 4. Composite Obfuscation Operation: S followed by E

Here, $P((x_u, y_u) \in C_{Init} \cap C_{Final})$ is the probability that the exact user location belongs to the intersection of the two circular regions; $P((x, y) \in C_{Init} \cap C_{Final}$ is the probability that a random location selected from the whole obfuscated region is within the intersection. In addition, (x_c, y_c) represents the original center and r is the original radius.

Finally, an example of the R obfuscation operator is shown in Figure 3(c). Here the initial radius r is reduced to $r' < r$. The analysis of this case is symmetric to that of the E operator, and we obtain the following relationship between the relevance of the argument and the result:

$$\mathcal{R}' = \frac{r'^2}{r^2} \cdot \mathcal{R}$$

Composite obfuscation is achieved by combining two operators. As operators E and R are inverse to each other, there are four kinds of composite operators: E followed by S, S followed by E, R followed by S, and S followed by R. An example of S followed by E is shown in Figure 4.

3.3   The iPDA Solution

Xu et al. [5] propose a client-based solution that enables privacy-preserving location-based data access, called iPDA. The basic idea behind iPDA is to transform a location-based query to a region-based query using an optimal location cloaking technique that is fully implemented on the client side. When a point-based query is transformed to a region-based query, the query result of the latter is a superset of that of the former. Xu et al. show that compared to any other shape with the same area, a circular enlarged region minimizes the size of the superset query result.

Mobility Analysis Attack

iPDA is designed to address the *mobility analysis attack*. Referring to Figure 5(a), a user first issues a query at location q, whose cloaking region is C_q. After a period

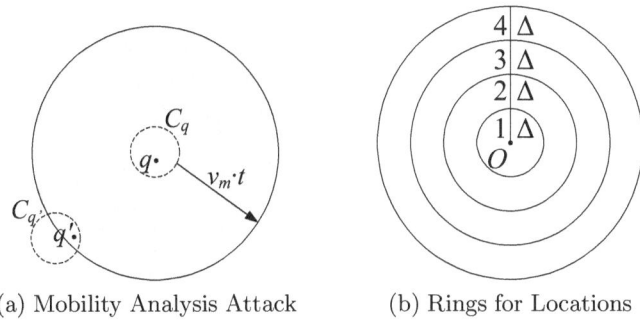

(a) Mobility Analysis Attack   (b) Rings for Locations

Fig. 5. iPDA Example

of time t, the user issues another query at location q' with a cloaking region $C_{q'}$. An adversary who knows the user's maximum speed v_m can infer that the user cannot be located within the white subregion of $C_{q'}$ when query q' is issued. Rather, the user must be within the intersection of $C_{q'}$ and the gray circular region expanded from C_q by the distance $v_m \cdot t$.

Privacy Model

iPDA is intended to generate cloaking regions for clients issuing queries repeatedly such that at any time a query is issued, the client's exact location uniformly distributed within the cloaking region. A general movement pattern is assumed to be known to both the client and the server. To carry out a numerical analysis, the plane of movement is divided into a set of rings with a common center O, as shown in Figure 5(b). Each ring, except the innermost one, which is actually a circular region, is of a sufficiently small width Δ.

Assume that at the time of the previous query, the cloaking region was centered at O and had radius $r = K \cdot \Delta$, and $R = L \cdot \Delta$ denotes the longest distance a user can travel between two queries. Both K and L are integers. Two probabilities can be defined based on the set of rings indicated by r and R.

The probability that the user's new location is in the i'th ring is captured by $U(i) = \int_{(i-1)\cdot\Delta}^{i\cdot\Delta} u(x)dx$, where $u(x)$ is the density function for the probability that the distance between the new user location and O is x. The conditional probability that the user's new location is in the i'th ring given that the center of the new cloaking region is in the j'th ring is denoted by $Q(i|j)$. Two expressions of $Q(i|j)$ are derived for the two cases $j < K$ and $j \geq K$ [5]. As a result, the location privacy against a mobility analysis attack is indicated by the value of $Q(i|j)$.

To enable a trade-off between query result accuracy and the cost of communication between client and server, queries are allowed to be blocked. A blocked query is not sent to the server; the result of the previous query is instead reused to derive the new result. If a blocked query comes from a ring $i \leq K$, the old result is a superset of the new result. This means that the query accuracy is not reduced while the communication cost is avoided. If a blocked query comes from a ring $i > K$, the old result is no longer a superset of the new result. This means that the communication cost is saved at the expense of the query accuracy. This also leads to a set of linear equations:

$$\sum_{j=max\{1,i-K+1\}}^{min\{L-K+1,i+K-1\}} \left(\frac{Q(i|j)}{U(i)} \cdot v_j\right) \leq 1 \quad i = 1, 2, \ldots, L$$

$$v_j = \sum_m (P(j|m) \cdot U(m)) \geq 0 \quad j = 1, 2, \ldots, L - K + 1$$

With these constraints, two linear programming techniques are developed to either maximize the query accuracy or minimize the query communication cost [5].

Query Processing

iPDA transforms a traditional kNN query to a query that requires the k nearest neighbors of a circle's perimeter Ω and the interior of Ω. Thus, a proposal for the server-side processing of k circular range nearest neighbor queries is also part of iPDA. A general heuristic is to access objects (or index nodes when the objects are indexed) in ascending order of their minimum distances to Ω. For disk-resident data, two pruning heuristics are proposed based on the distance and topology between Ω and index nodes.

Benefits and Limitations

Although the cloaking technique in iPDA is claimed to be optimal, discretization is needed before iPAD can be implemented using some numerical method. This reduces the attractiveness of iPAD. Nevertheless, iPDA addresses the mobility analysis attack and allows a user to issue consecutive queries while enjoying location privacy. Further, iPAD offers users the flexibility of either maximizing query accuracy or minimizing query communication cost when the original query is cloaked. As a last remark, a demonstration system is available that implements iPDA in a practical setting of a GPS-enabled Pocket PC and spatial database supported servers [6].

3.4   Server-Side Processing of Enlarged Queries

As range queries and k nearest neighbor queries are fundamental, they may be expected to be supported by location-based service servers. After applying query enlargement, an original location-based query is converted into a region-based query.

Region-Based Range Query

The starting point is a range query applied to an uncloaked user location. This query retrieves all data points within the range, which might be a circle or rectangle centered at the user location. The region-based range query occurs when the user location is cloaked by a region, again perhaps a circle or a rectangle. The region-based range query then returns all data points within the original range, but with any point in the cloaking region being the query point.

The resulting query can still be viewed as a range query, so the query is relatively straightforward to compute. For example, if the cloaking region is a (axis-aligned) rectangle and the query range is a (axis-aligned) rectangle, the region-based range query becomes rectangular (axis-aligned) range query.

Region-Based kNN Queries

The ensuing discussion assumes the nearest neighbor query (i.e., the special case with $k = 1$), but it is easily generalized to the k nearest neighbor query for any positive integer k.

An original point nearest neighbor query is transformed to a *range nearest neighbor query* (*RNN*) [15]. Given a range W, the *RNN* query returns the union of the nearest neighbors for each point in W, i.e., $RNN(W) = \{NN(p) \mid p \in W\}$. This definition implies that any object within W belongs to $RNN(W)$, as it is its own nearest neighbor. For any object o outside W, it holds [15] that:

$$o \in RNN(W) \Leftrightarrow \exists p \in \text{Border}(W)(o = NN(p))$$

Therefore, it suffices to compute $RNN(W)$ in two steps: (i) performing a range query with the region W, and (ii) computing the nearest neighbor of any point p located at the border of W.

In case the range W is a rectangle, the second operation can be further reduced into four *line segment based nearest neighbor* (LNN) queries [15] that can be evaluated by a continuous nearest neighbor (CNN) algorithm [16] that retrieves the nearest static points for any point on a given line segment. Hu and Lee [15] propose solutions for LNN queries on both memory-resident and disk-resident data.

An alternative server-side algorithm for processing the range nearest neighbor query is also available [17].

3.5   Location Privacy in Proximity-Based Services

Proximity-based services differ from services that rely on range and kNN queries. For example, a "friend finder" is a typical proximity-based service in which a user

A is alerted if a friend is within a specified distance δ_A of the user's current location. In this section, we introduce a privacy technique that employs location enlargement.

Setting

Mascetti et al. [7] propose techniques that offer location privacy in proximity based services. A service provideris assumed that receives (enlarged) user location updates, maintains (enlarged) user locations, and processes service requests from users by finding their nearby friends. The spatial domain of interest is abstracted as a *granularity* that consists of a number of non-overlapping *granules*. The granules are heterogenous in size and shape, and each is identified by an index.

For a given user, both the service provider and buddies (other users in the system) can be adversaries. Therefore, each user A needs to specify two granularities: G_A^{SP} defines the minimum location privacy requirement for the service provider, and G_A^U defines the requirement for the buddies. In either granularity, each granule is a minimum uncertain region.

SP-Filtering **Protocol**

Within the setting described above, the authors propose three privacy-aware communication protocols. We consider the *SP-Filtering* protocol, which is the only one that does not require peer-to-peer communication.

With this protocol, user A sends to a generalized location to the service provider when she updates her current location. Let A be located in the granule $G_A^{SP}(i)$, which is known by the service provider. The user's generalized location $L_A(i)$ is defined as follows:

$$L_A(i) = \bigcup_{i' \in \mathbb{N} \mid G_A^U(i') \cap G_A^{SP}(i) \neq \emptyset} G_A^U(i'),$$

which is the union of those G_A^U granules that intersects with $G_A^{SP}(i)$. Any other user B updates her location $L_B(j)$ similarly, where j is the index of the G_B^{SP} granule in which B is located.

Given two users A and B with generalized locations $L_A(i)$ and $L_B(j)$, the service provider determines the minimum distance $(d_{A,B})$ and maximum distance $(D_{A,B})$ between them. Based on $d_{A,B}$ and $D_{A,B}$, the service provider determines whether B is in the proximity of A. The three cases illustrated in Figure 6 occur. Here, where the actual locations are represented as dots while the generalized locations are rectangles.

If $D_{A,B} < \delta_A$, as shown in Figure 6(a), B must be in the proximity of A regardless of the exact locations of A and B are within their generalized location rectangles. In this case, the service provider informs A that "B is within proximity." If $d_{A,B} > \delta_A$, as shown in Figure 6(b), B has no chance to be within the proximity of A. In the last case, $d_{A,B} \leq \delta_A \leq D_{A,B}$, as shown in Figure 6(c). Here, it is uncertain whether B is within proximity of B. Thus, the service provider informs A that "B is possibly within proximity."

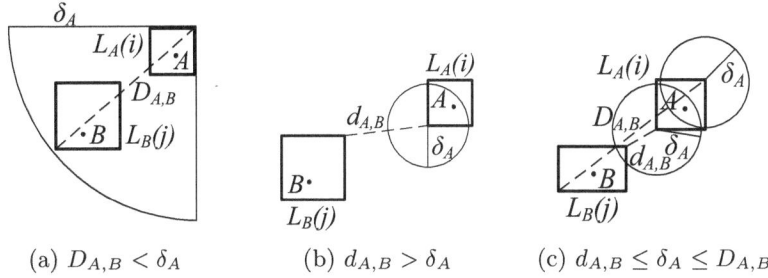

(a) $D_{A,B} < \delta_A$ (b) $d_{A,B} > \delta_A$ (c) $d_{A,B} \leq \delta_A \leq D_{A,B}$

Fig. 6. Example for SP-Filtering Protocol

Limitations and Extensions

The simple SP-Filtering protocol faces a dilemma. On the one hand, the granularity G_A^{SP} should be coarse for the purpose of location privacy. On the other hand, a coarse G_A^{SP} lowers the service accuracy, as the service provider is unable to determine whether buddy B is within proximity of A. Consequently, Mascetti et al. [7] propose two additional protocols, which, however, rely on peer-to-peer communication to derive a better result when an uncertain answer is returned.

4   Dummy-Based Techniques

We can regard the query enlargement techniques as *continuous*, in the sense that a user's location is enlarged into a closed region. In contrast, dummy-based techniques are *discrete* because the user's location is covered by multiple dummy locations, or dummies for short. All dummies, together with the user's location, are sent to the service provider, which is then unable to identify the user's real location. If a total of k locations are sent in a request, the service provider is then unable to identify the user's real location with a probability higher than $1/k$. This way, dummy-based techniques protect a user's location privacy.

We review two dummy-based location privacy protection techniques: a basic dummy-based technique [8] that offers limited control of the generation of dummies; and an enhanced dummy-based technique [9] that offers such controls and also takes into account the notion of privacy area to offer location privacy.

4.1   Basic Dummy-Based Technique

Format of a Message

In the basic dummy-based approach [8], a service request has the following format:

$$S = \langle u, L_1, L_2, \ldots, L_k \rangle$$

Here, u is a user identifier and $\langle L_1, L_2, \ldots, L_k \rangle$ is a set of locations consisting of the real user location and generated dummies. When the service provider receives a service request S, it processes the request for each location L_i ($1 \leq i \leq k$) according to the service type required, and then returns an answer R as follows:

$$R = \langle (L_1, D_1), (L_2, D_2), \ldots, (L_k, D_k) \rangle$$

Here, D_i is the services content for location L_i. When the user receives R, the service content D_r for the real query location L_r ($1 \leq r \leq k$) is selected as the result. Note that the real value of r is known only to the user.

Dummy Generation Algorithms
Given a region P, the service provider knows the (dummy) location cardinality in P as users send in requests. The location cardinality in P can vary from one time t to the next $t + 1$. If the cardinality difference for consecutive time points is too large, it is possible that the dummies move irregularly when compared with the user's actual movement. This, according to [8], causes the risk that adversaries may be able to identify user's real locations.

Thus, two dummy generation algorithms are proposed for users that issue service requests at each time step as they move [8]. In both algorithms, the first batch of dummy locations for a user are generated at random. The two algorithms then differ in how they generate subsequent dummy locations.

The *Moving in a Neighborhood* (MN) algorithm generates the next location of a dummy solely based on the current location of the dummy. Figure 7(a) illustrates the MN algorithm. It contains 11 dummies whose locations are represented by circles. The real user location L_r is drawn as a small dot. The arrow between each pair of locations indicates the movement during a time step. The new location of a dummy only depends on its previous location.

The *Moving in a Limited Neighborhood* (MLN) differs from the MN algorithm in that it takes into account the density of the region in which a newly generated dummy resides. Assuming that a user device is capable of obtaining the positions of other users, MLN regenerates a dummy if it finds that the dummy's generated location is in a region with too many users. Figure 7(b) gives an example of the MLN algorithm. During a time step, dummy L_i basically gets its new location

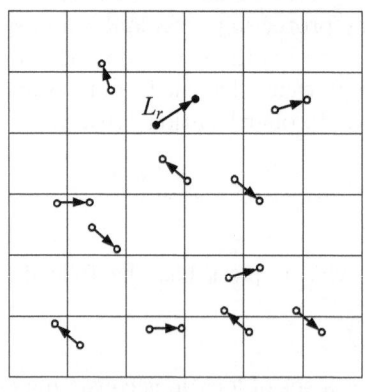

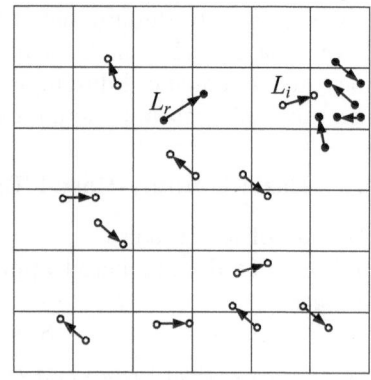

(a) MN Algorithm   (b) MLN Algorithm

Fig. 7. The MN and MLN Algorithms for Dummy Generation

according to the MN algorithm. However, the new location falls in a small range (a grid cell here) where there are four real locations of other users. Let the desired threshold of the region density be set at four. The MLN algorithm regenerates the new location for L_i at random until it is not in a dense region.

A drawback of the MLN algorithm is its reliance on the assumption that a user can obtain the real locations of nearby users. We remark that such a capability can itself cause location privacy issues.

Communication Cost Savings

In addition to dummy generation, communication cost reduction is also addressed [8]. As described, an original request is represented as $\langle u, (x_1, y_1), (x_2, y_2), \ldots, (x_k, y_k)\rangle$, where (x_r, y_r) $(1 \leq r \leq k)$ is the exact user location. Such a request ensures that the service provider is unable to identify the real user location with a probability higher than $1/k$. The request contains $8k$ bytes of location data if a single coordinate value takes 4 bytes (e.g., a float type value).

If all the x coordinates are put together followed by all the y coordinates in the request message, i.e., $\langle u, (x_1, x_2, \ldots, x_k), (y_1, y_2, \ldots, y_k)\rangle$, the message can be viewed as representing k^2 locations, namely the locations (x_i, y_j) $(1 \leq i, j \leq k)$. Thus, to ensure a probability of $1/k$, a request needs only use $2\sqrt{k}$ coordinates, or $8\sqrt{k}$ bytes of location data.

4.2   Enhanced Dummy-Based Technique

Although the MN and MLN algorithms are able to generate dummies that ensure that the service provider is unable to identify the real user location with a probability higher than $1/k$, they do not take into account the notion of the distances between the (dummy) locations. In particular, the region covered by all (dummy) locations, called the *privacy region*, is of importance because the area of that region indicates the difficulty for an adversary of tracking down the user. Neither MN nor MLN are capable of controlling the area of the privacy region.

Privacy Requirement

Motivated by this, Lu et al. [9] propose a privacy-area aware, dummy-based technique (PAD) for location privacy protection. PAD allows a user to specify privacy preference as $\langle k, s \rangle$, where k is the total number of locations in a request sent to the service provider and s is the area of the privacy region containing these k locations. Such a preference states that the service provider must be unable to identify the real user location with a probability higher than $1/k$ and must be unable to position the user in a region with area smaller than s.

Simply increasing the number of dummies in a request does not necessarily produce a larger privacy region. Therefore, new algorithms are needed to satisfy the privacy preference $\langle k, s \rangle$. Thus, two privacy-area aware dummy generation algorithms are proposed [9].

Circle-Based Dummy Generation

The circle-based dummy generation constrains all (dummy) locations, including the real user location, to a circle centered at position pos' with radius r.

Figure 8(a) shows an example where $k = 9$ and pos is the real user location. Each pair of clock-wise consecutive positions and pos' determines an angle θ. All positions are distributed in such a way that all θ values are equivalent.

The area $\tilde{s}$ of the hull of all positions is the sum of the areas of k triangles:

$$\tilde{s} = \sum_{i=1}^{k} \frac{1}{2} \cdot r_i \cdot r_{i+1} \cdot sin\theta = \frac{1}{2} \cdot \sum_{i=1}^{k} r_i \cdot r_{i+1} \cdot sin\frac{2\pi}{k},$$

where r_i is $dist(pos_i, pos')$. As there are only k locations in addition to pos', $pos_{k+1} = pos_1$ and $r_{k+1} = r_1$. Note that the hull is not necessarily convex and that $\tilde{s} \leq \hat{s}$, where $\hat{s}$ is the area of the corresponding convex hull. Assuming that all positions have identical distance to pos', the hull determined by them must be convex. Thus, taking into account the privacy area requirement s, the following holds:

$$\tilde{s} = \hat{s} = \frac{1}{2} \cdot k \cdot r_i^2 \cdot sin\frac{2\pi}{k} = s$$

Solving this produces an upper bound $r = \sqrt{(2 \cdot s)/(k \cdot sin\frac{2\pi}{k})}$. Let $r_{min} = \rho \cdot r$, where $0 < \rho \leq 1$. The following holds:

$$\hat{s} \geq \tilde{s} \geq \frac{1}{2} \cdot k \cdot r_{min}^2 \cdot sin\frac{2\pi}{k} = \frac{1}{2} \cdot k \cdot (\rho \cdot r)^2 \cdot sin\frac{2\pi}{k} = \rho^2 \cdot s$$

This indicates a lower bound of the privacy area of the k positions, i.e., $\hat{s} \geq \rho^2 \cdot s$. Thus, the virtual center pos' is determined at random such that $dist(pos, pos') \in [\rho \cdot r, r]$. As a result, by carefully choosing ρ, a guarantee can be gained on the privacy area of the location privacy query generated based on a virtual circle. For example, if we choose $\rho = \sqrt{3}/2$, we can ensure that the resulting privacy area is not smaller than three quarters of s.

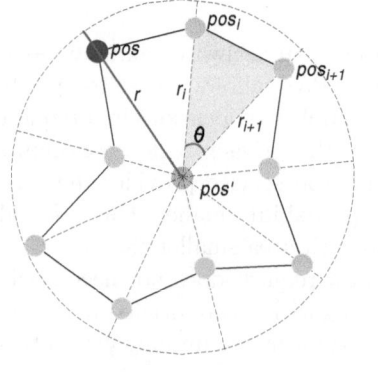

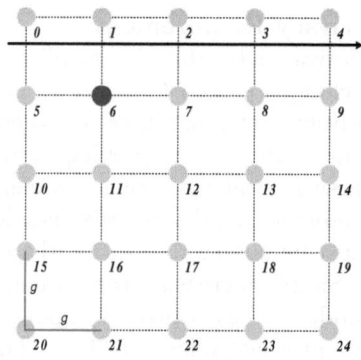

(a) Circle Based   (b) Grid Based

Fig. 8. Privacy-Area Aware Dummy Generation Examples

Grid-Based Dummy Generation

The circle-based algorithm aims to approximate the privacy area requirement s. In contrast, the grid-based algorithm always generates dummies whose privacy area is no smaller than the required s. The grid-based dummy generation works as follows. A (virtual) uniform, square grid is created, such that (i) it has k vertices, (ii) its area is equal to s, and (iii) the user position pos is one of the k vertices. The $k - 1$ other vertices are dummy locations, to be sent to the server together with the user position pos.

Figure 8(b) shows an example of the grid-based dummy generation, where $k = 25$ and 24 dummies are generated. All locations, including the real user location, are indexed in row-major order, and vertex 6 is the real user location. The side length of a grid cell is $g = \sqrt{s}/(\sqrt{k}-1)$. The coordinates of all dummies are determined by their indexes relative to the real user location.

When dummies are generated based on the virtual grid, the upstream communication cost can be further reduced compared to [8]. Instead of sending all coordinate values, it is possible to send the grid configuration only in the request. The configuration of a uniform grid is given by 3 parts: the top-left corner location (8 bytes), the side length of each square grid cell (4 bytes), and the number of grid cells in the horizontal/vertical direction (1 byte). Therefore, the location information in a request with the grid configuration consumes 13 bytes.

5   Progressive Retrieval Techniques

The solutions in this section are called as *progressive retrieval techniques* because they progressively retrieve potential result objects from the server until it is guaranteed that the exact result can be found or the user chooses to terminate the search with an approximate result.

5.1   Graph-Based Obfuscation

Duckham and Kulik [10, 11] study location privacy in the context of a graph, which is employed to model a road network. We first introduce their graph model and then elaborate on the procedure for processing a query.

Graph Model and Equivalence Class

In the graph model, each (possible) location l_i refers to a graph vertex. An edge between two locations l_i and l_j has an associated weight $w(l_i, l_j)$. The *network distance* $dist_N(l_i, l_j)$ between any two locations l_i and l_j is defined as the length of the shortest path between the two, i.e., the sum of the weights along the shortest path.

It is assumed that the data points (and the query object) are located at vertices. Figure 9a depicts a graph with seven locations $(l_1, l_2, \cdots, l_7)$. The numbers next to the edges indicate their weights. The dataset P contains two data points p_1 and p_2, which are located at l_7 and l_1, respectively. The network distance $dist_N(l_2, l_4)$, for example, is computed as $1 + 4 = 5$.

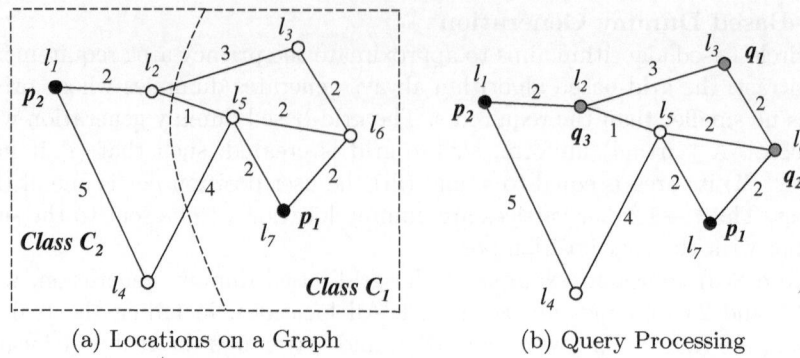

(a) Locations on a Graph   (b) Query Processing

Fig. 9. Obfuscation in a Graph Setting

By using the set P, the sets of locations can be partitioned into disjoint sets of *equivalence classes* such that each location in the same equivalence class has the same data point as its nearest neighbor. In Figure 9a, the locations l_3, l_5, l_6, and l_7 belong the same equivalence class (i.e., class C_1) because each of them takes p_1 as its nearest neighbor. The other locations belong to class C_2.

Obfuscation Set

Observe that the server maintains both the graph and the dataset P, whereas the user only knows the graph and her exact location q. Instead of submitting q to the server, the user needs to specify an *obfuscation set* Q, which is a set of graph vertices. The set Q is said to be *accurate* if it contains q, and it is said to be *imprecise* if it has more than one vertex [10]. The default setting is to choose a set Q that is both accurate and imprecise. Nevertheless, the user is allowed to choose a set Q that is inaccurate or precise. The consequence of using an inaccurate Q is that the actual query result is not guaranteed to be found. Intuitively, the user enjoys a high level of privacy when Q has a high cardinality or the user's exact location q is far from all the members of Q. However, it remains an open question how to combine both the accuracy and preciseness aspects into a unified, quantitative notion of privacy.

Negotiation Protocol for Query Processing

The user issues a nearest neighbor query by sending the obfuscation set Q to the server and following a *negotiation protocol*. If all vertices of Q belong to the same equivalence class (say, the class C_i), the server returns the data point of C_i as the result. Otherwise, the server negotiates with the user for a more precise obfuscation set Q' that is a proper subset of Q. If the user agrees to provide such a set Q', then thee protocol is applied recursively. If not, the server determines the largest equivalence class (say, the class C_j) that overlaps Q and returns the data points of C_j as the result.

We proceed to consider the query example in Figure 9b, with the obfuscation set $Q = \{q_1, q_2, q_3\}$. Since Q intersects with more than one equivalence class (i.e., classes C_1 and C_2), the server asks whether the user can provide a more precise obfuscation set Q'.

If the user prefers not to provide Q', the server checks whether the majority of the vertices of Q belong to the class C_1 or C_2. Since the majority of vertices of Q (i.e., q_1 and q_2) belong to class C_1, the server returns the point p_1 as the result.

In case the user accepts to provide a more precise set Q', say, $Q' = \{q_1, q_2\}$, the negotiation protocol is executed on Q' recursively. In this example, all vertices of Q' belong to the class C_1, so the server returns p_1 as the result.

Negotiation Strategies
Duckham and Kulik [10] also suggest strategies for the client to automate the negotiation process on the user's behalf. For example, their O-strategy reveals the equivalence class that covers the user upon the first negotiation; their C-strategy iteratively discards border locations from the obfuscation set; and their L-strategy provides the server with an inaccurate but precise location as the obfuscation set. In addition to these basic strategies, some advanced negotiation strategies are also discussed. Experimental results demonstrate that the O-strategy is able to achieve the best privacy protection [10].

Benefits and Limitations
As a remark, the above work is the first to study location privacy in the setting of a graph model. The negotiation protocol is a novel approach that enables the user to interactively control the trade-off between location privacy and query efficiency. From the viewpoint of user-friendliness, a user wishes to specify her desired privacy value without understanding the negotiation protocol and participating in negotiations. An open issue is to design a fully automatic negotiation policy for choosing the initial obfuscation set and the negotiation strategy.

The negotiation protocol has medium difficulty of implementation as it requires the server to compute the equivalence classes (of the dataset P) that intersect the user's obfuscation set Q (see Figure 9).

5.2   SpaceTwist

Yiu et al. [12] propose a client-based algorithm, called *SpaceTwist*, for retrieving the user's k nearest neighbors from the server without revealing the user's exact location q. In the following, we first describe the running steps of SpaceTwist, then examine its privacy model. At the end, we study the trade-offs among the privacy, performance, and result accuracy of SpaceTwist.

Query Execution of SpaceTwist
The server takes as input a "fake" location q' that differs from q. The location q' can be generated at the client side if the user specifies her exact location q and the distance $dist(q', q)$ between q' and q. For instance, the user sets $dist(q', q) =$ 500 m if she wants to obtain privacy at the level of a city block. In fact, the ability to vary $dist(q', q)$ enables a trade-off between location privacy and query efficiency. A location q' being far from q offers high privacy, but also leads to high query cost.

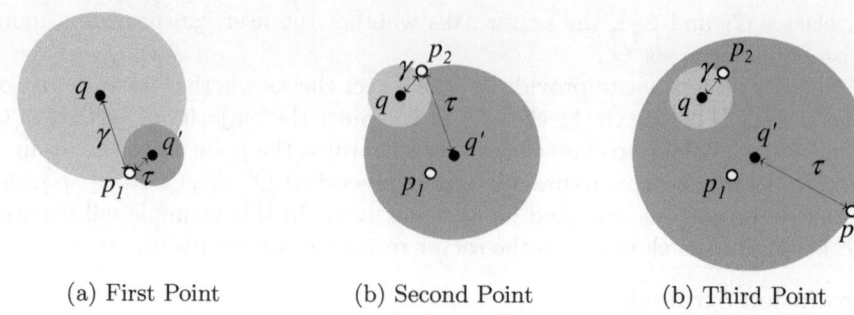

(a) First Point   (b) Second Point   (b) Third Point

Fig. 10. Query Processing in SpaceTwist

The algorithm then requests the server to retrieve data points in the ascending order of their distances from q'. This operation is known as incremental nearest neighbor retrieval [18], and it has been studied extensively in the literature. In order to reduce the number of communication packets, multiple points retrieved consecutively on the server are shipped to the client in a single packet. Let β be the number of points that can fit into a packet.

During execution, the algorithm maintains a result set W and two variables τ and γ. The variable τ represents the largest distance between any retrieved point and q' so far. The variable γ denotes the distance between q and its k k nearest neighbor, with respect to the set of points retrieved so far. Initially, W is the empty set, $\tau = 0$, and $\gamma = \infty$. Whenever a point p_i is retrieved, τ is updated to $dist(q', p_i)$. In case q is closer to p_i than some point in W, the algorithm updates both the set W and γ to reflect the best k nearest neighbors found so far. The algorithm guarantees that the actual k nearest neighbors are available to the user when the condition $\gamma + dist(q', q) \leq \tau$ is satisfied. When this condition is met, the algorithm terminates.

We proceed to illustrate the running steps of the SpaceTwist algorithm using the example in Figure 10. Assume that we have $k = 1$ and $\beta = 1$. After retrieving point p_1 (see Figure 10a), the best result is set to p_1. Both τ (dark gray circle) and γ (light gray circle) are updated. When point p_2 is retrieved (see Figure 10b), τ is updated. As q is closer to p_2 than the previous result (i.e., p_1), the best result becomes p_2 and γ is updated. Next, point p_3 is retrieved (see Figure 10c) and τ increases. Since $\gamma + dist(q', q) \leq \tau$ (i.e., the dark gray circle contains the light gray circle), the algorithm terminates the search on the server and returns p_2 as the nearest neighbor of q.

Privacy Model

The privacy study of the SpaceTwist algorithm [12] assumes that the adversary knows: (i) the point q' and the value k, (ii) the set of retrieved points from the server, and (iii) the termination condition of SpaceTwist. The goal of the adversary is to utilize the above information for determining whether a location q_c can be a *possible* user location. Note that q_c is not necessarily the same as the *actual* user location q.

Let m be the number of packets received by the client, and let their points (in their retrieval order) be $p_1, p_2, \cdots, p_{m\beta}$. It is shown that a possible user location q_c must satisfy both of the following inequalities [12]:

$$dist(q_c, q') + \min_{1 \leq i \leq (m-1)\beta}^{k} dist(q_c, p_i) > dist(q', p_{(m-1)\beta})$$

$$dist(q_c, q') + \min_{1 \leq i \leq m\beta}^{k} dist(q_c, p_i) \leq dist(q', p_{m\beta}),$$

where the term $\min_{1 \leq i \leq m\beta}^{k} dist(q_c, p_i)$ represents the distance between q_c and the kth nearest neighbor, with respect to the first $m \cdot \beta$ points retrieved.

The inferred privacy region Ψ is then defined as the set of all such possible locations q_c. Assume that both the point q' and the value k are fixed. It is worth noticing that the use of any location q_c in Ψ causes the SpaceTwist algorithm to retrieve the exact same sequence of points $(p_1, p_2, \cdots, p_{m\beta})$ as does the actual user location q. Thus, the adversary cannot observe the difference of q from the other points of Ψ based on the behavior of SpaceTwist.

The *privacy value* that quantifies the privacy obtained is then defined as the average distance between q and any point in Ψ:

$$\Upsilon(q, \Psi) = \frac{\int_{z \in \Psi} dist(z, q)\, dz}{\int_{z \in \Psi} dz}$$

Although the region Ψ can be inferred by both the user and the adversary, only the user can derive the privacy value $\Upsilon(q, \Psi)$ (which requires the knowledge of

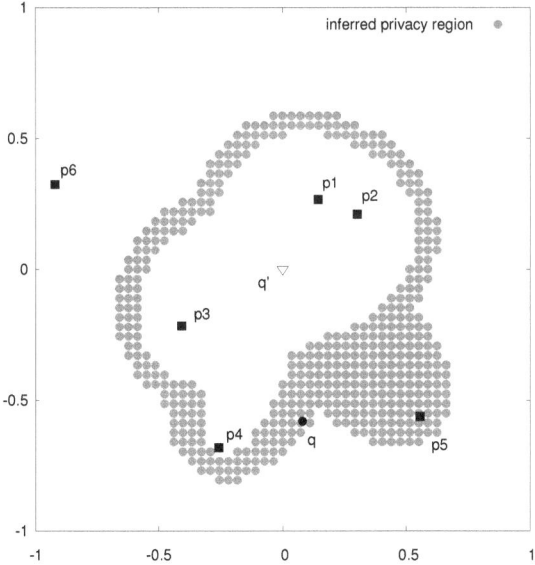

Fig. 11. Inferred Privacy Region

q). Figure 11 shows an example where the algorithm terminates after retrieving 6 data points from the server. The retrieved points $p_1, p_2, \cdots, p_6$ are labeled by their retrieval order. The nearest neighbor of q is the point p_4. The inferred privacy region corresponds to the gray region, which is an irregular ring that contains q.

Trade-offs among Location Privacy, Query Performance, and Query Accuracy

Yiu et al. [12] study a relaxed notion of the k nearest neighbors query. Given a query point q, a distance threshold ϵ, and a dataset P, the ϵ-relaxed k nearest neighbors query accepts a k-sized set W as the result if the maximum distance between q and W is upper-bounded by the sum of ϵ and the distance between q and the actual k nearest neighbor in P. The motivation of this approximate query is that a user (e.g., due to limited communication bandwidth) may be willing to accept a result that is not too far from her location q, if the cost can be reduced significantly.

A technique is then presented that computes the ϵ-relaxed k nearest neighbors query [12]. It employs a virtual grid structure to prune unnecessary points by utilizing the flexibility of the ϵ-relaxed k nearest neighbors query. It is shown that, the above technique guarantees the accuracies of the query results (within the ϵ bound), while saving communication cost (m) and improving the privacy value (Υ).

6   Transformation-Based Techniques

This section introduces *transformation-based* techniques that transform the original location-based query problem (e.g., range search, k nearest neighbors queries) into a search problem in another space.

6.1   Hilbert Curve Transformation

Transformation

Khoshgozaran ans Shahabi [13] propose to evaluate the k nearest neighbor query at an untrusted server by transforming all data points and the query point into one-dimensional numbers. This approach achieves complete privacy and constant query communication cost. However, it does not guarantee the accuracy of a query result.

The transformation function is implemented by a Hilbert curve function. The instance of the function being used is defined by an encryption key $\mathcal{EK}$ that consists of 5 values: a translation offset (X_t, Y_t), a rotation angle θ, a curve order O, and a scaling factor F. Let $\mathcal{H}(\mathcal{EK}, p_i)$ denote the Hilbert value of the data point p_i. An analysis suggests that there exists an exponential number of possible encryption keys and that it is infeasible for an adversary to infer the exact encryption key being used [13]. Without knowing the encryption key $\mathcal{EK}$, it is infeasible for the adversary to recover a data point p_i from its Hilbert value $\mathcal{H}(\mathcal{EK}, p_i)$.

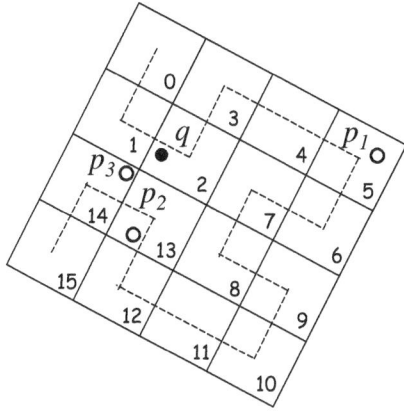

Fig. 12. Example of Hilbert Curve Transformation

Figure 12 depicts three data points p_1, p_2, and p_3 in the plane. The Hilbert curve is shown as a dashed curve, and each (square) cell is labeled with its Hilbert value. For example, the point p_1 is mapped to the value $\mathcal{H}(\mathcal{EK}, p_1) = 5$. Similarly, the values $\mathcal{H}(\mathcal{EK}, p_2) = 13$ and $\mathcal{H}(\mathcal{EK}, p_3) = 14$ are computed. These Hilbert values are then uploaded to the server. Observe that the server stores only three values (5, 13, 14), not the original data points.

Query Processing
At query time, the user applies the same encryption key to transform her location q into its Hilbert value $\mathcal{H}(\mathcal{EK}, q)$. Then the user requests the server to return k Hilbert values that are closest to the query Hilbert value. Next, the user applies the inverse transformation to obtain the result points from the retrieved Hilbert values.

We proceed to demonstrate an example of processing the $(k = 1)$ nearest neighbor query in Figure 12. At the client side, the query point q is mapped to the Hilbert value $\mathcal{H}(\mathcal{EK}, q) = 2$. Then the client asks the server to return the Hilbert value closest to 2. The server returns the value 5, the client decodes this back to the data point p_1, and it returns p_1 as the query result.

It is worth noticing that the retrieved point (say, p_1) is not necessarily the actual nearest neighbor of q (i.e., p_3). In fact, the solution does not provide any guarantee on the retrieved point.

6.2   Private Information Retrieval

Ghinita et al. [14] propose a solution for private nearest neighbor search by applying a *computationally private information retrieval protocol*. The core idea is that the client can efficiently test whether a large number is quadratic residue (QR) or quadratic non-residue (QNR); however, the adversary (e.g., the service provider) cannot efficiently do so. Under modulo arithmetic, QR and QNR are analogous to a bit value 0 and 1, respectively.

In the pre-processing phase, the Voronoi cell $V(p_i)$ of each data point p_i is computed. Then the domain space is partitioned into a regular grid of $G \times G$ cells, and each grid cell c_j stores an m-bitmap to represent the data points whose Voronoi cells intersect the spatial region of c_j.

At query time, the client first locates the cell c_q that contains the user location q. It then computes a sequence of G large numbers such that the number corresponding the column of c_q is a QNR and the others are QR. The server then performs modulo arithmetic on the above sequence of numbers by using the bit values in the grid. Essentially, the server obtains a sequence of G numbers, corresponding to the content of cells at the same column as c_q. The client then picks the number in the retrieved sequence that is in the same row as c_q. If that number is QNR, the original bit value in that cell is 0; otherwise, it is 1. This procedure is repeated for each of the m bit positions in order to obtain the complete content (of data points) stored in c_q.

This solution always returns the actual query results, and it is proven to be computationally secure (i.e., achieving perfect privacy) [14]. The solution is specially designed for the nearest neighbor query; its extension to the k nearest neighbors query has not been studied. A drawback of the solution is that it incurs high execution time on the server-side, limiting the server from being concurrently used by massive amount users. Empirical studies show that it takes 20 seconds to process the exact nearest neighbor query using a single-CPU server [14].

7 Promising Directions and Open Problems

Location privacy solutions for the client-server architecture are of high interest in the sense that this architecture is simple and widely deployed. While good advances have been made, several general directions for future research exist.

First, it is possible to extend techniques covered in this chapter to 2-dimensional space with obstacles, which are regions where service users cannot be located. Such obstacles usually are known to the server or adversaries, who can make use of this information to increase their success of guessing the real user location. For example, the dummy generation algorithms should not use any locations within those obstacles as dummies. For query enlargement techniques, the intersection between an enlarged query and the obstacles should be minimized, or taken into account, to ensure location privacy. One possibility would be to build on the formal model of bettini et al. [2] for obfuscation-based techniques.

Second, it is relevant to extend the techniques covered in this chapter to simultaneously support Euclidean space (with obstacles) and road network space. For instance, it is relevant to consider the extension of some works [10, 11] to also apply to Euclidean space, and the extensions of other works [3, 4, 5, 6, 8, 9, 12, 13, 14] to also apply to road network space.

Third, while most techniques covered in this chapter consider only snapshot queries, it is of interest to offer more proposals to support also continuous queries that can be issued by mobile users. For query enlargement techniques, it is of interest to find more efficient server-side query processing approaches for enlarged queries.

Fourth, transformation-based techniques covered [13, 14] are promising in terms of the privacy they offer. However, they are not readily deployable in existing settings. A recent approach [19], though not designed for location privacy applications, devises a keyed transformation function such that its output domain remains to be the 2-dimensional space. This way, existing spatial indexes and query processing techniques are leveraged for the query processing. Unfortunately, this approach is applicable to range queries only, and it assumes that all the clients share the same key value. It remains an open problem to extend this approach to become a solution to the query location privacy problem.

Last but not least, there is a need for a location privacy solution that can be accepted by both the clients and the service provider. In practice, the service provider may be interested in certain aggregate statistics over the whole population of users (e.g., "finding the region with the highest density of users at 5 p.m."), rather than tracking the clients' exact locations. It is challenging to develop a solution that support the above aggregate query and yet satisfies the users' privacy requirements.

Acknowledgments

This work was conducted in part while Christian S. Jensen was on sabbatical at Google Inc. from his position at Aalborg University.

References

1. Voelcker, J.: Stalked by Satellite: An Alarming Rise in GPS-enabled Harassment. IEEE Spectrum 47(7), 15–16 (2006)
2. Bettini, C., Mascetti, S., Wang, X.S., Jajodia, S.: Anonymity in Location-Based Services: Towards a General Framework. In: MDM, pp. 69–76 (2007)
3. Cheng, R., Zhang, Y., Bertino, E., Prabhakar, S.: Preserving User Location Privacy in Mobile Data Management Infrastructures. In: Privacy Enhancing Technology Workshop, pp. 393–412 (2006)
4. Ardagna, C.A., Cremonini, M., Damiani, E., di Vimercati, S.D.C., Samarati, P.: Location Privacy Protection Through Obfuscation-Based Techniques. In: DBSec, pp. 47–60 (2007)
5. Xu, J., Du, J., Tang, X., Hu, H.: Privacy-Preserving Location-based Queries in Mobile Environments. Technical Report, Hong Kong Baptist University (2006)
6. Du, J., Xu, J., Tang, X., Hu, H.: iPDA: Supporting Privacy-Preserving Location-Based Mobile Services. In: MDM, pp. 212–214 (2007)
7. Mascetti, S., Bettini, C., Freni, D., Wang, X.S., Jajodia, S.: Privacy-aware proximity based services. In: MDM, pp. 1140–1143 (2009)
8. Kido, H., Yanagisawa, Y., Satoh, T.: An Anonymous Communication Technique using Dummies for Location-based Services. In: IEEE International Conference on Pervasive Services (ICPS), pp. 88–97 (2005)

9. Lu, H., Jensen, C.S., Yiu, M.L.: PAD: Privacy-Area Aware, Dummy-Based Location Privacy in Mobile Services. In: MobiDE, pp. 16–23 (2008)
10. Duckham, M., Kulik, L.: Simulation of Obfuscation and Negotiation for Location Privacy. In: Cohn, A.G., Mark, D.M. (eds.) COSIT 2005. LNCS, vol. 3693, pp. 31–48. Springer, Heidelberg (2005)
11. Duckham, M., Kulik, L.: A Formal Model of Obfuscation and Negotiation for Location Privacy. In: Gellersen, H.-W., Want, R., Schmidt, A. (eds.) PERVASIVE 2005. LNCS, vol. 3468, pp. 152–170. Springer, Heidelberg (2005)
12. Yiu, M.L., Jensen, C.S., Huang, X., Lu, H.: SpaceTwist: Managing the Trade-Offs Among Location Privacy, Query Performance, and Query Accuracy in Mobile Services. In: ICDE, pp. 366–375 (2008)
13. Khoshgozaran, A., Shahabi, C.: Blind Evaluation of Nearest Neighbor Queries Using Space Transformation to Preserve Location Privacy. In: Papadias, D., Zhang, D., Kollios, G. (eds.) SSTD 2007. LNCS, vol. 4605, pp. 239–257. Springer, Heidelberg (2007)
14. Ghinita, G., Kalnis, P., Khoshgozaran, A., Shahabi, C., Tan, K.L.: Private Queries in Location Based Services: Anonymizers are not Necessary. In: SIGMOD, pp. 121–132 (2008)
15. Hu, H., Lee, D.L.: Range Nearest-Neighbor Query. IEEE TKDE 18(1), 78–91 (2006)
16. Tao, Y., Papadias, D., Shen, Q.: Continuous Nearest Neighbor Search. In: VLDB, pp. 287–298 (2002)
17. Mokbel, M.F., Chow, C.Y., Aref, W.G.: The New Casper: Query Processing for Location Services without Compromising Privacy. In: VLDB, pp. 763–774 (2006)
18. Hjaltason, G.R., Samet, H.: Distance Browsing in Spatial Databases. TODS 24(2), 265–318 (1999)
19. Yiu, M.L., Ghinita, G., Jensen, C.S., Kalnis, P.: Outsourcing Search Services on Private Spatial Data. In: ICDE, pp. 1140–1143 (2009)

Private Information Retrieval Techniques for Enabling Location Privacy in Location-Based Services*

Ali Khoshgozaran and Cyrus Shahabi

University of Southern California
Department of Computer Science
Information Laboratory (InfoLab)
Los Angeles, CA 90089-0781
{jafkhosh,shahabi}@usc.edu

Abstract. The ubiquity of smartphones and other location-aware hand-held devices has resulted in a dramatic increase in popularity of location-based services (LBS) tailored to user locations. The comfort of LBS comes with a privacy cost. Various distressing privacy violations caused by sharing sensitive location information with potentially malicious services have highlighted the importance of location privacy research aiming to protect user privacy while interacting with LBS.

The anonymity and cloaking-based approaches proposed to address this problem cannot provide stringent privacy guarantees without incurring costly computation and communication overhead. Furthermore, they mostly require a trusted intermediate anonymizer to protect a user's location information during query processing. In this chapter, we review a set of fundamental approaches based on *private information retrieval* to process range and k-nearest neighbor queries, the elemental queries used in many Location Based Services, with significantly stronger privacy guarantees as opposed to cloaking or anonymity approaches.

1 Introduction

The increasing availability of handheld computing devices and their ubiquity have resulted in an explosive growth of services tailored to a user's location. Users subscribe to these *location-aware* services and form spatial queries (such as range or k-nearest neighbor search) to enquire about the location of nearby points of interest (POI) such as gas stations, restaurants and hospitals. Processing these queries requires information about the location of the query point or a query

---

* This research has been funded in part by NSF grants IIS-0238560 (PECASE), IIS-0534761, IIS-0742811 and CNS-0831505 (CyberTrust), and in part from the METRANS Transportation Center, under grants from USDOT and Caltrans. Any opinions, findings, and conclusions or recommendations expressed in this material are those of the author(s) and do not necessarily reflect the views of the National Science Foundation.

C. Bettini et al. (Eds.): Privacy in Location-Based Applications, LNCS 5599, pp. 59–83, 2009.
© Springer-Verlag Berlin Heidelberg 2009

region of interest. However, providing this information to a potentially untrusted location-based server has serious privacy implications as it can easily reveal the querying user's location information. Misusing this sensitive information as well as other malpractices in handling such data have resulted in a variety of distressing and increasingly more concerning privacy violations.

Similar to many other existing approaches in areas such as data mining and databases, various techniques based on the K-anonymity principle [1] have been extensively used to provide location privacy [2,3,4,5,6,7]. With these approaches, usually a trusted third party known as the *anonymizer* is used to ensure that the probability of identifying the querying user remains under $\frac{1}{K}$ where K is the size of the anonymity set received by the untrusted location server. Alternatively, users can generate the anonymity set in a decentralized fashion. With these approaches, the user's location is usually *cloaked* in a larger region which includes other users to make it harder for the untrusted server to locate the querying user. Aside from requiring users to trust a third party during query processing (or all other users for the decentralized case), recent studies [8,9,10,11] have shown that such approaches suffer from many drawbacks such as an insufficient guarantee of perfect privacy, vulnerability to correlation attacks and a huge performance hit for privacy paranoid users. To overcome such restrictions, a new class of transformation-based techniques have emerged that map a user's location to a space unknown to the untrusted server. Using the query transformation process, the untrusted server is blinded while processing spatial queries to ensure location privacy [8,12]. Although these approaches mitigate some of the privacy implications of the anonymity and cloaking-based approaches, they cannot provide strong privacy guarantees against more sophisticated adversaries. Finally, several cryptographic-based approaches are proposed for location privacy which utilize two-party computation schemes to achieve privacy [13,14,15]. While these approaches can provide strong privacy guarantees, they suffer from yet another drawback. They cannot avoid a linear scan of the entire data and thus are not efficient for real-world scenarios.

In this chapter, we review two fundamental approaches that go beyond the conventional approaches proposed for location privacy and devise frameworks to eliminate the need for an anonymizer in location-based services and satisfy significantly more stringent privacy guarantees as compared to the anonymity/cloaking-based approaches. Both of these techniques are based on the theory of

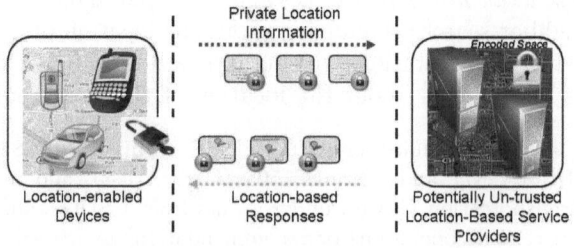

Fig. 1. Location Privacy in LBS

Private Information Retrieval (PIR) to protect sensitive information regarding user locations from malicious entities. Using a PIR protocol, a client can retrieve a database item hosted at an untrusted server without revealing which item is retrieved from the host (Figure 1). Although PIR can be used to privately generate a query result set, avoiding a linear private scan of the entire object space is challenging. This is due to the fact that the server owning the objects information cannot be trusted to perform the query processing and choose what to be returned as responses. Alternatively, moving this knowledge to the users will require the query processing to happen at the client side which is very costly. Utilizing spatial partitionings based on PIR, these approaches devise private algorithms that significantly reduce the amount of information that is privately transferred to the querying clients from the untrusted server. While the first approach relies on hardware-based PIR techniques [10,16], the second one employs computational PIR protocols to provide location privacy [9]. We elaborate on each approach and detail the merits and shortcomings of both techniques.

The remainder of this chapter is organized as follows. In Section 2, we review some related work. Section 3 details various PIR schemes and in Section 4, we first provide our trust and threat model and then show how PIR can be used to design privacy-aware spatial query processing. Sections 5 and 6 elaborate on how hardware-based and computational PIR techniques are employed to enable private evaluation of spatial queries, respectively. In Section 7, we discuss some of the limitations of PIR-based approaches and finally, Section 8 concludes the book chapter with possible future research directions.

2   A Brief Survey

Protecting a user's private location information while interacting with location-based services has been the subject of many recent research efforts. These studies can be broken into four fundamentally different groups of anonymity/cloaking, transformation, cryptographic and PIR-based approaches.

Anonymity and Cloaking-Based Approaches: The earlier work on location privacy focused on protecting a user's private location information by disguising it among $K-1$ other user locations or extending it from a *point* location to an area (spatial extent). With the first approach, user u, as well as $K-1$ other user locations form an *anonymity set* which is sent to the server instead of u's precise location. Similarly, with cloaking techniques, the resulting *cloaked* region (which contains u and several other users) is sent to the server. These techniques try to ensure the user's location cannot be distinguished from the location of the other $K-1$ users or the exact user location within the cloaked region is not revealed to the untrusted server responding to location queries. Depending on the method used, the untrusted server executes the query for every object in the anonymity set or for the entire cloaked region. Several techniques based on cloaking and K-anonymity have been proposed in the literature to reduce the probability of identifying a user's location [2,3,4,5,6,7].

Cloaking and K-anonymity approaches have some important limitations. First, by design the majority of cloaking approaches rely on a trusted intermediary to "anonymize" user locations which means all queries should involve the *anonymizer* during the system's normal mode of operation. The anonymization can also be performed in a decentralized fashion among users which means each user has to trust all other users in the system with her location. In other words, while users do not trust the location server, they either have to trust another third party, as sophisticated as the server or all other users. Second, a limitation of cloaking techniques in general is that either the quality of service or overall system performance degrades significantly as users choose to have more strict privacy preferences. Third, many of the cloaking techniques are subject to attacks that exploit the information from the formation of the cloaked region or the history of user movement to *infer* precise user location [9]. Fourth, the concept of K-anonymity does not work in all scenarios. For example, in a less populated area, the size of the extended area can be prohibitively large in order to include $K-1$ other users. Even worse, not enough number of users may be subscribed to the service to construct the required cloaked region. Finally, these techniques assume that all users are trustworthy. However, if some of them are malicious, they can easily collude to compromise the privacy of a targeted user. The interested reader might refer to the "Anonymity and Historical-Anonymity in Location-Based Services" chapter for a more detailed description of the anonymity and cloaking techniques.

Transformation-Based Approaches: A second class of approaches emerging to mitigate some of the weaknesses of the techniques discussed above are based on query transformation to prevent the server from learning information about user locations. The first work to utilize spatial transformation techniques for location privacy is [8]. In this study, space filling curves are utilized as one-way transformations to encode the locations of both users and points of interest into an encrypted space and to evaluate a query in this transformed space. The transformed space maintains the distance properties of the original space which enables efficient evaluation of location queries in the transformed space. Subsequently, upon receiving transformed query results, users can reverse the transformation efficiently using the trapdoor information which is only provided to them and are protected from the server. Recently, Yiu et al. proposed a framework termed SpaceTwist to blind an untrusted location server by incrementally retrieving points of interest based on their ascending distance from a fake location near the query point termed the anchor point [12]. Note that with this approach, the query is still evaluated in the original space but the query point is transformed to an anchor point.

The key advantage of transformation-based approaches over the anonymity and cloaking-based techniques is the elimination of the need for a trusted third party during the query processing. Furthermore, [8] provides very efficient query processing without compromising privacy and [12] utilizes the existing query processing index structures present in non-privacy aware servers which makes it readily applied to existing location servers. However, the nearest neighbor algorithm of [8] is approximate and [12] suffers from several privacy leaks and costly computation/communication if exact results and strict privacy are required simultaneously. Furthermore, it offers no lower bound for the size of the privacy region where it can become even smaller than a cloaked region.

Cryptographic-Based Approaches: This class of techniques blind the untrusted party (i.e., the server or another user) by utilizing secure multi-party computation schemes. The protocol proposed in [14] privately evaluates the distance between Alice's point and other n points that Bob owns. After executing the protocol, Bob knows nothing about Alice's point and Alice only learns the nearest neighbor from Bob's points. Although the solution proposed is mainly of theoretical interest and does not focus on spatial queries or LBS, it can be considered as a method for providing location privacy in LBS. Zhong, Goldberg and Hengartner propose three solutions to what they define as the "nearby-friend problem" [15]. The problem is defined as allowing users to learn information about their friends' locations if and only if their friends are actually nearby. The three protocols are all efficient in terms of the amount of computation and communication required by each party. Each protocol is an instance of a multi-party computation scheme with certain strengths and restrictions (in terms of number of messages transferred and the resilience to a malicious party). Finally, Zhong et al. provide two protocols aiming at protecting user locations in LBS. While the first protocol allows a user to share her location information with other users via an untrusted server, the second protocol enables a dating service where a user learns whether other users with similar profiles (found by the server) are located in the same region she is located [13]. This protocol, which is of more interest to location privacy in LBS, assumes the entire user profile is known to the server, and the server first finds any potential matches between a user and all other users. The server then sends all matched profiles to the requester so that she can blindly compare their locations with her own location. Similar to the other protocols discussed above, a multi-party computation protocol is proposed which involves the requester, the dating service and any other matched user.

The main advantage of the three methods discussed above is their strong privacy guarantees. Building their framework on well-known cryptographic primitives and widely used one-way functions, these protocols do not suffer from privacy leaks of cloaking/anonymity and transformation-based methods. Furthermore, their problem-specific designs allow very efficient implementations of the protocols mostly involving only a handful of computations and few message transfers. However, the fundamental disadvantage of the protocols discussed in this category is their high computation or communication complexity when being used for spatial query processing. For instance, in [14] the distance between query point and each and every point of interest must both be computed or transferred to the client, i.e., $O(n)$ computation or communication complexity where n is the size of the database. This is because the points of interest are treated as vectors with no exploitation of the fact that they are in fact points in space. Therefore, the main limitation of cryptographic-based techniques discussed above is the loss of spatial information via encryption. This loss either results in a linear scan of the entire database if used to evaluate a spatial query (as in [14]), or makes the protocol unusable for spatial query processing (as in [15,13]). Similarly, with the protocols proposed in [15], Alice will know whether a certain user Bob is nearby. However, verifying whether a certain friend is nearby Alice is a different problem than finding Alice's nearest friends. Finally, the work of Zhong et al. suffers from the same drawbacks since it only describes a matching protocol in an encrypted space between two users located in the same region. However, the real challenge is finding nearby matches which is not possible in an encrypted space.

PIR-Based Approaches: Several approaches discussed so far attempt to improve the efficiency or the privacy aspects of evaluating spatial queries privately in LBS. However, they mostly suffer from a privacy/quality of service trade-off. While on one extreme end the cryptographic-based techniques provide perfect privacy, they result in very costly spatial query processing schemes. Likewise, on the other side of the spectrum, efficient cloaking or spatial transformation approaches might result in severe privacy leaks under certain user, object or query distributions. The approaches studied in this category are based on the solutions proposed to the well-known problem of *Private Information Retrieval* (PIR) discussed in Section 1. These approaches construct private spatial indexes on top of PIR operations to provide efficient spatial query processing, while the underlying PIR scheme guarantees privacy.

In this chapter, we discuss two location privacy schemes based on hardware-based [10,16] and computational PIR [9] protocols. The former approach superimposes a regular grid on the data and uses PIR to privately evaluate range and kNN queries. The latter technique supports approximate and exact nearest neighbor query evaluation by utilizing various 1-D and 2-D partitionings to index the data and then restructuring partitions into a matrix that can be privately queried using PIR. Sections 5 and 6 detail these two approaches and Section 7 compares the two techniques and highlights the strengths and weaknesses of each approach. To the best of our knowledge, the only other study to propose employing PIR to provide location privacy is [11] which presents an architecture that uses PIR and trusted computing to hide location information from an untrusted server. With this approach, PIR is used to prevent the untrusted location server from learning user locations and trusted computing is used to ensure users that the PIR algorithm and other services provided by the server are only performing the operations as intended. In fact, similar to hardware-based PIR, [11] places a trusted module as close as possible to the untrusted host to disguise the selection of records. However, the proposed techniques do not specifically focus on spatial query processing (such as range and kNN) and the proposed architecture is not yet implemented.

In summary, the novel approaches discussed in this chapter do not rely on anonymizers for query processing. Furthermore, they do not suffer from the privacy vulnerabilities of cloaking/anonymity and transformation-based approaches or the prohibitive communication and computation costs of cryptographic-based techniques. In the next section, we formally define the PIR problem and detail some of its implementations.

3   Private Information Retrieval

Private Information Retrieval, in its most abstract setting, enables a user to query an item from a database without disclosing the queried item to the (potentially) untrusted server hosting the data. Suppose Alice owns a database D of n bits and Bob is interested to retrieve the ith bit from the database. PIR protocols allow Bob to privately retrieve $D[i]$ without disclosing i to Alice. This definition of PIR offers a theoretical point of view. In a more practical scheme, users are interested in privately retrieving blocks of data (or records) [17].

Given the privacy requirements of users, PIR approaches can be divided based on whether they provide *Information Theoretic* or *Computational* privacy. While the former class of approaches guarantee privacy against an adversary with unbounded computational power, the latter class assumes a computationally bounded adversary. Therefore, the information theoretic approaches guarantee perfect privacy while the security of the computational approaches relies on the intractability of a computationally complex mathematical problem, such as Quadratic Residuosity Assumption [18]. However, the perfect privacy of the first group comes with a prohibitive cost. In fact, Chor et al. have proved the communication cost of such techniques to be $\Omega(n)$ for a database of n bits [17]. Furthermore, the server's computation cost is also linear since not processing any single database record r indicates to the server that r is not requested by the user and thus by definition violates the privacy requirement of PIR. Therefore, while being of theoretical interest, information theoretic PIR cannot efficiently be integrated into data-intensive and practical applications. The computational PIR approaches, on the other hand, achieve significantly lower complexity by assuming some limitations on the server's computational power.

While computational PIR incurs more reasonable costs for retrieving objects, the proposed PIR protocols are still expensive and require a significant amount of server resources. In other words, although they can improve the communication complexity, all database records still have to be processed at the server. In fact, Sion et al. argue in [19] that the cost of privately retrieving database items from the server is significantly higher than sending the entire database to the client. This argument has certain important restrictions [9] and we show that practical results can be achieved by avoiding some of the redundant PIR operation costs. However, the per item cost of computational PIR approaches are still high.

To obtain perfect privacy while avoiding the high cost of the approaches discussed above, a new class of *Hardware-based PIR* approaches has recently emerged which places the trust on a tamper-resistant hardware device. These techniques benefit from highly efficient computations at the cost of relying on a hardware device to provide privacy [20,21,22,23]. Placing a trusted module very close to the untrusted host allows these techniques to achieve optimal computation and communication cost compared to the computational PIR approaches.

In this chapter, we show how recent studies have used these two classes of PIR approaches to enable location privacy for range and kNN queries. We stress that there are various other versions of the PIR problem that we do not consider in this chapter. For instance, the PIR techniques we have discussed so far are all instances of *single server* PIR schemes. A different set of studies focus on multi-server PIR protocols. Here, it is assumed that several servers exist to execute the PIR operations with the requirement that these servers should not be able to communicate or collaborate with each other. Under this assumption of non-communicating servers, it is possible to achieve sub-linear communication complexity [17,24,25]. However, the underlying non-collusion assumptions made by these studies are rather hard to achieve and makes it difficult to develop practical schemes based on these multi-server PIR protocols. In another setting, *symmetric PIR* strives to protect server privacy as well as user privacy by limiting user's knowledge to the physical value of the retrieved item [26]. Such schemes are not of particular interest in the context of location-based services as we assume server data is publicly available and thus server privacy is not as important as user privacy. We detail our trust and threat model assumptions in Section 4.1.

4   PIR in Location Privacy

With many location-based services, users carrying location-aware handheld devices are interested in finding the location of nearby points of interest (POI) such as restaurants and hotels. Users form various spatial queries such as range or kNN queries to request such information. Therefore, the location of the query point (or region), as well as the query result set usually reveal the location of the user. The key idea behind using PIR techniques for location privacy is to prevent the untrusted location server from learning any information about a query and its result set. Using PIR, users can request information about their locations of interest without revealing any information about their whereabouts. However, a major challenge in performing this task is that users are essentially unaware of the way records are indexed on the untrusted server and hence cannot directly request the records that might contain their desired information. Therefore, avoiding a linear scan or full transfer of the entire server database is challenging. This is due to the fact that the server owning the objects information cannot be trusted to perform the query processing and choose what to be queried. Alternatively, moving this knowledge to users will require the query processing to happen at the client side which is very costly. For the rest of this section, we discuss how private PIR-based spatial algorithms address this issue. We first define the trust and the threat models and then detail how spatial queries are translated to PIR requests.

4.1   Trust and Threat Model

We consider a model in which users query a central untrusted server for POI data. While users trust their client devices to run legitimate software, they do not trust any other entity in the system including the location server (henceforth denoted by LS). Users might collude with LS against other users and thus from each user's point of view, all other users as well as LS can be adversarial. LS owns and maintains a database of POIs and responds to users queries as a service provider. Users subscribe to LS's services. As part of our threat model, we assume that the server's database is publicly accessible and available and thus an adversary can perform the so-called *known plaintext attack*.

As we discussed earlier, an adversary's goal is to find a user's location information. Therefore, the obvious objective of any location privacy scheme is to protect such private information from potentially malicious servers and other adversaries. In order to achieve location privacy, a user's location and identity information, as well as the identity of query results should be kept secret both on the server and during query evaluation [8].

We assume there is a secure communication channel between users and LS and thus the connection cannot be sniffed by adversaries. However, the server can gain valuable information from user queries as well as their result sets and therefore, these entities should not leak any information to an adversary. Based on our assumption of a secure client-server communication channel, no adversary can learn about a user's location without colluding with the server. Therefore, for the rest of this chapter, we only focus on the location server as the most powerful adversary and assume that adversaries are computationally bounded.

4.2   Converting Spatial Queries to PIR Requests

One important property of the PIR problem is the underlying assumption on the protocol usage; it is assumed that the bits (or records) are stored in an array and users know the index of the element they wish to retrieve. Therefore, the key problem is to enable users to privately map location queries into their corresponding record indexes of the database.

With this problem characteristic in mind, it is clear that the key step in utilizing PIR for spatial query processing is to devise efficient schemes which allow users to find objects relevant to their queries that should be privately retrieved from a remote database. In this section, we elaborate this argument and discuss several techniques to utilize PIR for location privacy. In particular, we study the recent work that addresses private evaluation of range, NN and kNN queries. In Section 5, we discuss how hardware-based PIR techniques are employed to enable location privacy. Similarly, Section 6 presents an approach to privately evaluate nearest neighbor queries using computational PIR.

5   Location Privacy with Hardware-Based PIR Protocol

The class of hardware-based PIR techniques utilize a secure coprocessor to disguise the selection of records that are requested by the user from an untrusted server. The secure coprocessor performs certain operations that prevent the server from learning the requested record from database items read by the coprocessor. In this section, we review how a secure coprocessor is used to implement a PIR protocol to privately and efficiently retrieve a selected record from a database.

5.1 Hardware-Based PIR

A Secure Coprocessor (SC) is a general purpose computer designed to meet rigorous security requirements that assure unobservable and unmolested running of the code residing on it even in the physical presence of an adversary [23]. These devices are equipped with hardware cryptographic accelerators that enable efficient implementation of cryptographic algorithms such as DES and RSA [19].

Secure coprocessors have been successfully used in various real-world applications such as data mining [27] and trusted co-servers for Apache web-server security [28] where the server hosting the data is not trusted. The idea behind using a secure coprocessor for performing the PIR operations is to place a trusted entity as close as possible to the untrusted host to disguise the selection of desired records within a black box.

Placing a secure coprocessor between user queries and the untrusted server raises the following simple yet important question. Why should one not trust a location server if the secure coprocessor is to be trusted? The response to this question is based on several fundamental differences between trusting a secure processor versus a location server. First, aside from being built as a tamper resistant device, the secure coprocessor is a hardware device specifically programmed to perform a given task while a location server consists of a variety of applications using a shared memory. Secondly, unlike the secure coprocessor in which the users only have to trust the designer, using a location server requires users to trust the server admin and all applications running on it as well as its designer. Last but not least, in our setting, the secure coprocessor is mainly a *computing* device that receives its necessary information, per session from the server, as opposed to a server that both *stores* location information and *processes* spatial queries.

We build our location privacy scheme based on the PIR protocols proposed in [20,21] to achieve optimal (i.e., constant) query computation and communication complexity at the cost of performing as much offline precomputation as possible. We now provide an overview of our utilized PIR protocol.

Definition 1. Random Permutation: For a database DB of n items the random permutation π transforms DB into DB_π such that $DB[i] = DB_\pi[\pi[i]]$. For example for $DB = \{o_1, o_2, o_3\}$ and $DB_\pi = \{o_3, o_1, o_2\}$ the permutation π represents the mapping $\pi = \{2, 3, 1\}$. Therefore $DB[1] = DB_\pi[\pi[1]] = DB_\pi[2] = o_1$, $DB[3] = DB_\pi[\pi[3]] = DB_\pi[1] = o_3$, etc. It is easy to verify that the minimum space required to store a permutation π of n records is $n \log n$ bits.

In order to implement the PIR protocol, we first use the secure coprocessor to privately shuffle and encrypt the items of the entire dataset DB using a random permutation π. Once the shuffling is performed, DB_π is written back to the server while SC keeps π for itself. To process a query $q = DB[i]$, a user u encrypts q using SC's public key and sends it to SC through a secure channel. SC decrypts q to find i and retrieves $DB_\pi[\pi[i]]$ from the server, decrypts and then re-encrypts it with u's public key and sends it back to u (hereinafter we distinguish between a *queried item* which is the item requested by the user and *retrieved/read record* which is the item SC reads from DB_π). However, the above scheme is not yet private as (not) retrieving the same object for the second query reveals to the server that the two queries are (different) identical. Therefore, SC has to maintain a list L of all items retrieved so far. SC also caches the records retrieved from the beginning of each session. In order to answer the kth query, SC first searches its cache. If the item does not exist in its cache, SC retrieves $DB_\pi[\pi[k]]$ and stores it in its cache. However, if the element is already cached, it randomly reads a record not present in its cache and caches it. With this approach, each record of the database might be read at most once regardless of what items are queried by users. This way, an adversary monitoring the database reads can obtain no information about the record being retrieved. The problem with this approach is that after $T_{threshold}$ retrievals, SC's cache becomes full. At this time a *reshuffling* is performed on DB_π to clear the cache and L. Note that since $T_{threshold}$ is a constant number independent of n, query computation and communication cost remain constant if several instances of reshuffled datasets are created offline [21], alternatively shuffling can be performed regularly on the fly which makes the query processing complexity equal to the complexity of the shuffling performed regularly.

Algorithm 1. *read(permuted database DB_π, index i)*

Require: DB_π, T {Threshold}, L {Retrieved Items}
1: **if** $(|L| \geq T)$ **then**
2:     $DB_\pi \leftarrow$ Reshuffle DB_π using a new random permutation π;
3:     $L \leftarrow \emptyset$;
4:     Clear SC's cache
5: **end if**
6: **if** $i \notin L$ **then**
7:     $record \leftarrow DB_\pi[\pi[i]]$;
8:     Add $record$ to SC's cache
9:     $L = L \cup \{i\}$;
10: **else**
11:     $r \leftarrow$ random index from $DB_\pi \setminus L$;
12:     $temp \leftarrow DB_\pi[\pi[r]]$;
13:     Add $temp$ to SC's cache
14:     $L = L \cup \{r\}$;
15: **end if**
16: return $record$;

Algorithm 1 illustrates the details of the *read* operation which privately retrieves an element from the database. Note that it reads a different record per query and thus ensures each record is accessed at most once. All details regarding the shuffling, reshuffling, permutation etc. are hidden from the entity interacting with DB_π.

So far we have enabled private retrieval from an untrusted server. However, we have not focused on how spatial queries can be evaluated privately. Section 5.1 enables replacing a normal database in a conventional query processing with its privacy-aware variant. However, the query processing needs to be able to utilize this new privacy-aware database as well. Note that what distinguishes our work from the use of encrypted databases is the impossibility of blindly evaluating a sophisticated spatial query on an encrypted database without a linear scan of all encrypted items. In this section, we propose private index structures that enable blind evaluation of spatial queries efficiently and privately. Using these index structures, we devise a sweeping algorithm to process range queries in Section 5.2. Similarly, we detail a Spiral and a Hilbert-based approach (or Hilbert for short) to privately evaluate kNN queries in Section 5.3.

5.2   Private Range Queries

As we discussed in Section 1, the key idea behind employing spatial index structures is to avoid the private retrieval of database objects not relevant to user queries. However, a challenge in designing these indexes raises from the fact that while they are stored at the untrusted host, query processing cannot be performed by the server. This requirement forces us to use efficient indexes that can quickly identify the subset of database records that should be privately retrieved. For processing range (and kNN) queries, we utilize a regular $\delta \times \delta$ grid to index objects within each cell. The key reason behind using a grid structure in our framework is while being efficient, grids simplify the query processing. Several studies have shown the significant efficiency of using the grid structure for evaluating range, kNN and other types of spatial queries [29,30,31].

Without loss of generality, we assume the entire area enclosing all objects is represented by a unit square. The grid index uniformly partitions the unit square into cells with side length δ ($0 < \delta < 1$). Each cell is identified by its cell ID (c_{id}) and may contain several objects each being represented by the triplet $< x_i, y_i, obj_{id} >$. These cells are then used to construct the $listDB$ index which stores the objects and their location information for each cell. The $listDB$ schema represents a flat grid and looks like $< c_{id}, list >$ where $list$ is a sequence of triplets representing objects falling in each grid cell. Figure 2 illustrates the original object space and the $listDB$ index. There is an obvious trade-off between two competing factors in choosing the right value of δ. As δ grows, a coarser grid (having less cells) decreases the total number of cell retrievals during query processings. This is desirable given the relatively high cost of each private read from the database. However, large cells result in retrieving more excessive (unneeded) objects which coexist in the cell being retrieved. These excessive objects result in

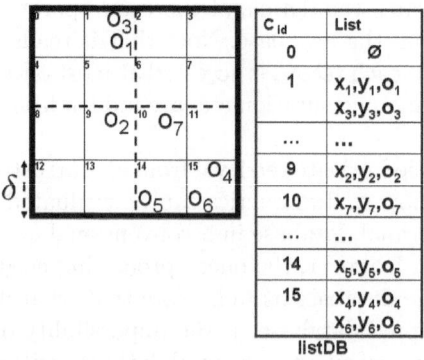

Fig. 2. The Object Space (left) and listDB Index (right)

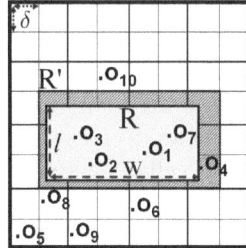

Fig. 3. Range Query Processing

higher computation and communication complexity which increase the overall response time. These trade-offs and the discussion on choosing the right grid granularity are studied in more detail in [10,16].

Using the *listDB* index, processing range queries is straightforward. During the offline process, database records are created each containing a *listDB* entry which corresponds to all objects within a cell. To ensure that the server cannot distinguish between the records based on differences in record sizes (which is affected by object distributions), each record is padded to reach the maximum record length. Next, SC generates a random permutation π and privately shuffles the and encrypts *listDB*. The encrypted shuffled version of *listDB* is then written back to the server. During the range query processing, SC uses a sweeping algorithm to privately query the server for all cells which overlap with the specified range. A range query $range(R)$ is defined as a rectangle of size $l \times w$ ($0 < l, w \leq 1$). Therefore, to answer each range query using *listDB*, we must first find the set of cells R' that encloses R. R' forms a $L \times W$ rectangular grid where $L \leq \lceil \frac{l}{\delta} \rceil + 1$ and $W \leq \lceil \frac{w}{\delta} \rceil + 1$. The function $read(listDB, c_{id})$ privately queries *listDB* and performs the necessary processing to return a list of all objects enclosed in a c_{id}. We use the sweeping algorithm to query the cells in R' privately.

5.3   Private kNN Queries

The main challenge in evaluating kNN queries originates from the fact that the distribution of points can affect the size of the region R that contains the result set (and hence the cells that should be retrieved). In other words, no region is guaranteed to contain the k nearest objects to a query point (except in a uniform distribution) which implies that R has to be progressively computed based on object distributions. Therefore, it is important to minimize the total number of cells that should be privately queried. In this Section, we propose two variants of evaluating kNN queries and discuss how our index structures allow us to query only a small subset of the entire object space. Note that due to the strong similarity of these algorithms with their *first nearest neighbor* counterparts (i.e., $k = 1$), we directly consider the more general case of kNN.

Similar to range query processing, we can use regular grids and the *listDB* index to perform kNN queries. However, despite its simplicity, *listDB* might not be an efficient index for skewed object distributions as the kNN algorithm utilizing *listDB* might experience a performance degradation caused by numerous empty cells in the grid. Increasing δ does not solve this problem as it results in coarse-grained cells containing many excessive objects that have to be queried/processed and even a linear decrease in δ incurs at least a quadratic increase in the number of empty cells. Therefore, we also introduce a *hilbDB* index which uses Hilbert space filling curves [32] to avoid the stated shortcomings of processing kNN queries using regular grids. The main intuition behind using Hilbert curves is to use their locality preserving properties to efficiently approximate the nearest objects to a query point by only indexing and querying the non-empty cells. This property significantly reduces the query response time for skewed datasets [16].

We define H_2^N ($N \geq 1$), the Nth order Hilbert curve in a 2-dimensional space, as a linear ordering which maps an integer set $[0, 2^{2N} - 1]$ into a 2-dimensional integer space $[0, 2^N - 1]^2$ defined as $H = \nu(P)$ for $H \in [0, 2^{2N} - 1]$, where P is the coordinate of each point. The output of this function is denoted by *H-value*.

To create the *hilbDB* index, an H_2^N Hilbert curve is constructed traversing the entire space. After visiting each cell C, its $c_{id} = \nu(C)$ is computed. We use an efficient bitwise interleaving algorithm from [33] to compute the H-values (the cost of performing this operation is $O(n)$ where n is the number of bits required to represent a Hilbert value). Next, similar to the *listDB* index, the c_{id} values are used to store object information for each cell. Finally, in order to guide the next retrieval, each record also keeps the index of its non-empty c_{id} neighbors in *hilbDB*, stored in the *Prev* and *Next* columns, respectively. These two values allow us to find out which cell to query next from *hilbDB* hosted at the server. Figure 4 illustrates the original object space and the *hilbDB* index. The circled numbers denote each cell's c_{id} constructed by H_2^2 for the *hilbDB* index. For clarity, we have not shown that all records are in fact encrypted.

Using the two private indexes discussed above, the following general approach is employed by both of our kNN algorithms: *(i)* create a region R and set it to the cell containing the query point q *(ii)* expand R until it encloses at least k

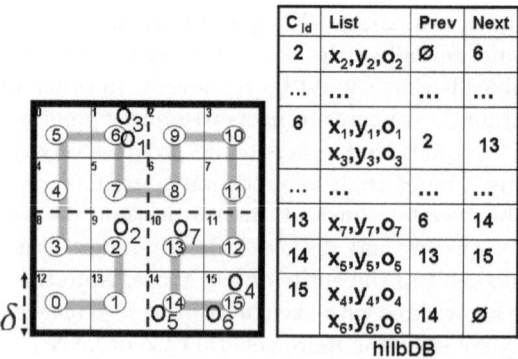

Fig. 4. The Hilbert Curve Ordering of Objects (left) and the hilbDB Index (right)

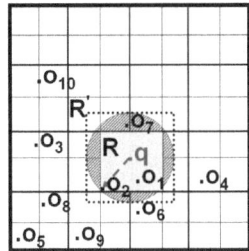

Fig. 5. Computing the Safe Region R'

objects *(iii)* compute the *safe region* R' as the region guaranteed to enclose the result set and *(iv)* find the actual k nearest objects in R' using $range(R')$ defined in Section 5.2. The main difference among the two algorithms is related to how they perform the step *(ii)* mentioned above. Note that, as Figure 5 illustrates, regardless of the approach, R is not guaranteed to contain the actual k nearest neighbors of q and therefore has to be expanded to the safe region R' (e.g., $O_7 \in 2NN(q)$ but $O_7 \notin R$ and $O_2 \in R$ although $O_2 \notin 2NN(q)$). As shown, if relative location of q and its furthest neighbor in R is known, a safe region can be constructed. However, this comes at the cost of querying *listDB* to retrieve the location of objects in R. This is in fact useful since $R \subset R'$ which means R has to be retrieved sometime during the query processing. Therefore, as an optimization, by querying every cell in R during the safe region computation, we can avoid querying them again for $range(R')$. It is easy to verify that R' is a square with sides $2 \times \lceil \|c_q - far_q(k)\| \rceil$ where c_q is the cell containing q and $far_q(k)$ is the cell containing q's k^{th} nearest object in R and $\|.\|$ is the Euclidean norm [31]. We now elaborate on how different expansion strategies for step *(ii)* mentioned above generate different results and discuss the pros and cons of each strategy.

Spiral Expansion. With this approach, if k objects are not found in the cell containing the query point, we expand the region R in a spiral pattern until it encloses k objects. The most important advantage of this method is its simple and conservative expansion strategy. This property guarantees that the spiral expansion minimizes R. However, the very same conservative strategy might also become its drawback. This is because for non-uniform datasets, it takes more time until the algorithm reaches a valid R. Note that the Spiral Expansion only uses the *listDB* index to evaluate a kNN query. Figure 6 (left) illustrates the order in which grid cells are examined.

Hilbert Expansion. The spiral expansion strategy, and in general similar linear expansion strategies such as the one proposed in [31], are usually very efficient in finding the safe region for the query q due to their simplicity. However, the time it takes to find the region that includes at least k objects can be prohibitively large for certain datasets. The Hilbert expansion overcomes this drawback by navigating in the *hilbDB* index which only stores cells that include at least one object. The main advantage of using *hilbDB* is that once the cell with closest

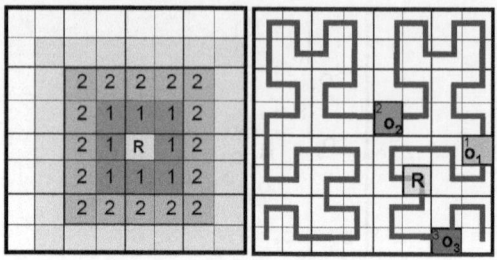

Fig. 6. kNN Query Processing

H-value to $\nu(q)$ is found, expanding the search in either direction requires only $k - 1$ more private reads to generate a safe region. This 1-dimensional search gives a huge performance gain at the cost of generating a larger search region due to the asymmetric and 1-dimensional Hilbert expansion. These trade-offs are extensively studied in [16]. Figure 6 (right) shows how cells containing objects are examined based on their Hilbert ordering.

6   Location Privacy with Computational PIR Protocol

To avoid the impractical linear lower bound on communication complexity of information theoretic PIR protocols, the privacy requirements can be relaxed from information theoretic to computational secrecy. This latter class of approaches known as computational PIR enforces computational intractability for any adversary to find the item being queried, from the client server communications. We now briefly describe the protocol and its underlying *Quadratic Residuacity Assumption (QRA)* [18] which acts as the basis for the secrecy of the computational PIR protocol.

6.1   Computational PIR

In its theoretical setting, a computational PIR protocol allows a client to query for the value of the ith bit from a server's database D of n bits that enables the client to derive the value of $D[i]$ while preventing the server to learn i. According to the QRA assumption, it is computationally hard to determine the quadratic and non-quadratic residues in modulo arithmetic for a large number N without knowing its two large prime factors q_1 and q_2. However, knowing q_1, q_2, one can efficiently determine the quadratic residues and quadratic non-residues modulo N. More details about the QRA assumption and its PIR usage can be found in [9]. Suppose the user wants to retrieve $D[i]$. Using this property, the server first converts D into a square $t \times t$ matrix M (which can be padded for non-square matrixes). Suppose that $M_{a,b}$ represents the requested value. The user first generates the modulus N using q_1, q_2 and then creates a row vector query message $y = [y_1, \ldots, y_t]$ such that only y_b is a quadratic non-residue and the rest of the vector are quadratic residues. The server computes a response z_r for each matrix row r according to equations 1,2 and sends z back to the user.

$$z = [z_1, \ldots, z_t] \text{ s.t. } z_r = \prod_{j=1}^{t} w_{r,j} \qquad (1)$$

$$w_{r,j} = y_j^2 \text{ if } M_{r,j} = 0 \text{ or } y_j \text{ otherwise} \qquad (2)$$

Using the Euler criterion and the Legendre symbol [34], the user can efficiently determine whether $M_{a,b} = 0$ or not [9]. Figure 7 illustrates these steps. Note that the above procedure allows the user to privately retrieve a single bit from the server. For objects represented as $m-$bit binary strings, the server can conceptually maintain m matrixes each storing one bit of each object and applying user's vector query message to all m matrixes. Similar to the PIR protocol described in Section 5.1, we assume $read(D_i)$ privately retrieves the ith object from D using the above procedure. Equations 1 and 2 as well as the above discussion can be used to compute the PIR protocol complexity. For each PIR read, server's computation is linear in n and the client sever communication is $O(\sqrt{n})$. This is because the entire matrix is multiplied by the user's query vector and the server response is a column vector of size $O(\sqrt{n})$. Similarly, for objects of size m bits, computation and communication complexities increase to $O(m.n)$ and $O(m\sqrt{n})$, respectively.

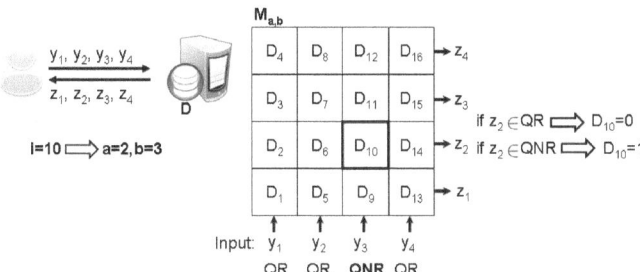

Fig. 7. Computational PIR

From the description of the protocol, it is clear that in order to successfully employ the computational PIR for a database D of n bits, data has to be first transformed into its matrix representation. In other words, D is first translated into a $\lceil \sqrt{n} \rceil \times \lceil \sqrt{n} \rceil$ matrix. Therefore, any private request from the server retrieves $O(\sqrt{n})$ bits to the user. The challenge is then how to organize data into buckets of size $O(\sqrt{n})$ in order to maximize query result accuracy. Note that this requirement is imposed by the employed PIR scheme and is independent of the underlying indexing scheme used. Sections 6.2 and 6.3, detail several underlying spatial indexing techniques proposed in [9] that enable private evaluation of approximate and exact nearest neighbor queries, respectively.

6.2   Approximate Private Nearest Neighbor Queries

Querying server's data privately with the computational PIR scheme requires some preprocessing to convert the original data into its PIR matrix representation. However, the challenge is to devise efficient index structures on top of the underlying PIR model to efficiently and privately respond to nearest neighbor queries. Once the data is bucketized into $\sqrt{n}$ pieces offline, PIR can be used to retrieve the section most relevant to the user's query. In this section, we review several 1-D and 2-D data partitioning schemes originally proposed in [9] that use various spatial index structures such as Hilbert curves, kd-trees and R-trees to privately evaluate NN queries.

Hilbert Curves: As we mentioned in Section 5.3, the major property of Hilbert curves is the proximity of objects in the Hilbert space. Using these curves, during the offline preprocessing step, the corresponding Hilbert values are computed for all POIs in the database D. Given a sorted list of these n Hilbert values, binary search can be used to find the nearest POI in terms of its Hilbert distance to the Hilbert value of the query point q in $O(\lg n)$ steps (Figure 8). This point is most probably the closest point to q. However, the logarithmic cost of each PIR request makes it impractical to use the above technique during the query processing. More importantly, each PIR request includes not one but $O(\lg n)$ points in its response. These factors lead to a very costly communication cost of $O(\sqrt{n}lgn)$ POIs for each approximate nearest neighbor search. For instance, finding the nearest neighbor to a point in a database of one million POIs would require the server to transfer approximately $20K$ points to the client while the result might still not be exact.

To mitigate the prohibitive communication cost of a single PIR operation, POIs can be indexed using a B^+-tree of height 2 where each node contains less than $\lceil \sqrt{n} \rceil$ points. The nodes of the B^+-tree represent the columns of the PIR matrix M. Given a query point q, the user u first computes $r = \nu(q)$ which denotes the Hilbert value associated to the query point. All values in a tree leaf are less or equal to their corresponding key stored at the tree root. Therefore, given the root, r is used to determine the tree node (corresponding to a matrix column) that should be retrieved. The user u then computes his nearest neighbor

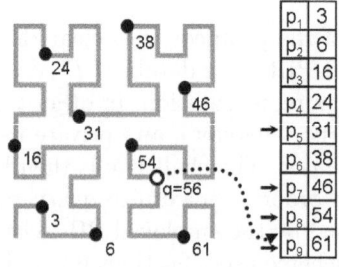

Fig. 8. Hilbert Values of POI Data

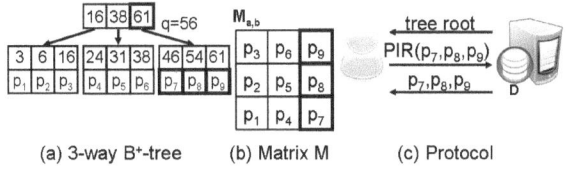

Fig. 9. 1-D Nearest Neighbor Approximation

from the $\sqrt{n}$ POIs returned in the result set (Figure 9). Although retrieving several POIs at the vicinity of q can greatly improve accuracy, the method remains an approximation technique. Note that any information required by all queries does not need to be retrieved privately. In our case, the tree root is required for processing any query and thus its plain value is first transferred to u. Since the query is evaluated with one PIR request, the communication complexity is $O(\sqrt{n})$. Figure 9 illustrates how the objects in Figure 8 are organized into the 3-way B^+-tree. For $r = 56 = \nu(q)$, u requests the third column of M since $56 > 38$. Next, u computes $NN(q)$ from the result set $\{p_7, p_8, p_9\}$.

kd-Trees and r-Trees: In the previous approach, dimension reduction techniques are used to retrieve POIs in the neighborhood of the querying point. However, any other spatial indexing scheme can employ the underlying PIR protocol as long as the data is effectively partitioned into buckets of size $\sqrt{n}$. Here, we briefly explain how the previous approach can be extended with 2-D partitionings such as kd-trees and r-trees [9]. Similar to the above case, the idea is to partition nearby POIs into buckets of size at most $\sqrt{n}$. This partitioning guarantees that we end up with at most $\sqrt{n}$ buckets hence conforming to the PIR matrix requirements.

For the case of kd-trees, the original algorithm partitions data horizontally or vertically until each partition holds a single point. The tree construction is modified such that the space is recursively partitioned into most balanced eligible splits. In other words, at each step, the algorithm recursively chooses the most balanced split among the remaining partitions provided that the total number of partitions does not exceed $\sqrt{n}$. This partitioning is illustrated in Figure 10a.

Similar to kd-trees, the r-tree construction algorithm has to be modified in order to guarantee the PIR matrix requirements as follows. The new construction scheme does not allow a root node of more than $\sqrt{n}$ MBRs. Therefore, the recursive r-tree partitioning algorithm is modified to ensure that all MBRs contain equal number of POIs. The query processing for r-trees is similar to the kd-tree case. Figure 10b illustrates the r-tree partitioning. Figure 10 also illustrates how a block of POI data is privately retrieved from the server under each partitioning to find the nearest neighbor of the query point.

From the discussion above, it is obvious that all three algorithms retrieve $O(\sqrt{n})$ POIs for each nearest neighbor query (which can be used to return the approximate k^{th} nearest neighbor where $1 \leq k \leq \sqrt{n}$). Furthermore, it provides a general strategy for utilizing any spatial partitioning for private evaluation of nearest neighbor queries. Therefore, the choice of partitioning strategy only

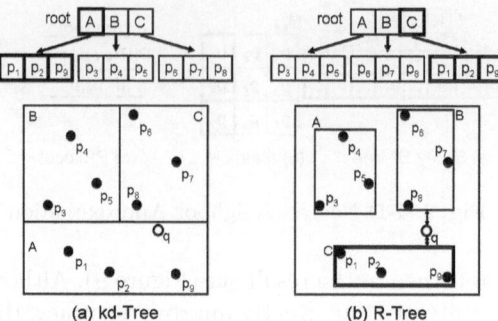

Fig. 10. 2-D Nearest Neighbor Approximations

determines the approximation error and does not increase the communication complexity. More details regarding the empirical comparison of these strategies can be found in [9].

6.3 Exact Private Nearest Neighbor Queries

Providing exact answers to nearest neighbor queries, require spatial index structures and algorithms that guarantee the exact answer is always returned to the user as part of the result set. Voronoi diagrams [35] have an important property which makes them very suitable for this purpose. If each POI p_i is used as the generator of a Voronoi cell, by definition p_i is the nearest neighbor of any point within that cell. Using this property, the Voronoi tessellation of all POIs is first computed. Next a regular $\delta \times \delta$ grid is superimposed on top of the Voronoi diagram where each grid cell C holds the generator for each Voronoi cell with which it intersects[1]. This assignment guarantees that by returning all Voronoi cell generators corresponding to a grid cell C, the exact nearest neighbor to any query point q within C is included in the result set.

During the query processing, the user u first learns the grid granularity much in the same way he learns the tree roots in Section 6.2. Next, u finds the grid cell C which includes the query point q and privately requests for the content of C. As we discussed above, the result is guaranteed to include u's nearest neighbor. Note that the PIR matrix is employed in a different fashion compared to the approximate methods. First, while the entire result set is used to improve the approximation error in the case of approximate nearest neighbor search, the redundant POI data associated with a matrix column returned to the user has no use as the algorithm by construction guarantees the exact result to be included in the content of the grid cell enclosing q. Second, instead of individual POIs, all POIs within grid cells form the elements of the PIR matrix. Therefore, different grid granularities and object distributions might result in records of different sizes which in turn leads to an information leak while employing the PIR protocol. This is because the uneven object size distributions allow the attacker to

---

[1] This approach can be extended to support range queries by storing the set of POIs each cell encloses.

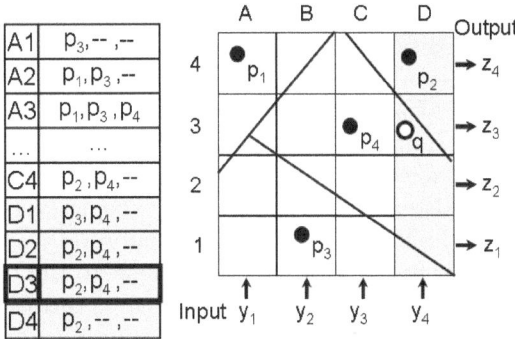

Fig. 11. Exact Nearest Neighbor Search

distinguish between different PIR records. To avoid this vulnerability, all cells are padded with dummy data based on the density of the most crowded grid cell. Figure 11 illustrates the padding and the exact nearest neighbor search process. Since u's query q is located in $D3$, it privately computes p_4 as its nearest neighbor. The empirical comparison between the exact and approximate nearest neighbor search strategies discussed above as well as the efficiency of each partitioning algorithm are presented in [9].

7 Open Problems

As we discussed in Sections 1 and 2, PIR-based approaches to location privacy are the first to provide perfect secrecy as well as provable security against correlation attacks while evaluating spatial queries. Furthermore, these approaches are not vulnerable to an adversary's prior knowledge about object distributions as the object retrieval patterns reveal no information to a malicious entity. These strong measures of privacy give a clear advantage to PIR-based techniques compared to other approaches for location privacy. However, the stringent privacy guarantees does not come without a cost.

With computational-based PIR, the server computation time as well as the communication complexity are rather high. Several techniques can be used to improve these complexities. For instance, compression techniques can utilize significant amount of redundancy in server's response to a PIR request and result in a 90% improvement in communication cost. As another optimization, we note that the PIR matrix M does not have to be square and changing its shape can improve the communication cost without exacerbating server's computation cost. Finally, data mining techniques can be used to reduce the redundancy in the multiplications of the PIR matrix to a user's request vector. This can also save up to 40% of server's computation cost[2]. However, none of these optimizations can break the linear server computation restriction which is inevitable by the definition of computational PIR. While empirical evaluations show that the

---

[2] These optimizations are detailed in [9].

implementation of the PIR operations and the query processing algorithms result in practical privacy-aware location-based services, the server's response time is still relatively high (several seconds) [9].

The class of hardware-based PIR approaches, on the other hand, avoid several inefficiencies of the computational PIR schemes by providing optimal server computation and communication costs. However, these optimizations are due to reliance on a tamper-resistant secure coprocessor. The major drawback of secure coprocessors are their limited storage and computation resources. Therefore, even very simple algorithms take significantly longer to run on a secure coprocessor compared to their execution time on non-secure processing platforms. The security measures used in securely packaging the processing unit imposes significant thermal and space restrictions. These factors highly limit the potential of embedding faster processors inside the secure package. Furthermore, from a financial standpoint, the relative high cost of secure coprocessors makes it harder to justify their use for some real-world scenarios.

From a practical point of view, PIR problem characteristics limit the efficient processing of spatial queries. As discussed in Section 4.2, computational PIR-based approaches to location privacy require a certain formulation of the problem which converts spatial query processing to private retrievals from a matrix representation of server's data. Similarly, hardware-based PIR approaches require frequent shuffling of data performed by the relatively slow secure coprocessor. These factors make it challenging to efficiently process spatial queries where the main objective is retrieving as few database items as possible for a given query.

Finally, another important restriction in the context of privacy-aware spatial query processing of PIR-based techniques is their inability to efficiently cope with location updates. The PIR-based techniques we discussed so far assume the location data is static and hence the data can be prepared for PIR operations during an offline process. However, with many emerging location-based and social networking scenarios, users are not merely requesting information about the POI around them but would also like to share their presence with their buddies and query their buddies' current locations as well. This requirement (i.e., querying dynamic location data) poses a fundamental challenge to PIR approaches. One opportunity for computational PIR schemes is to employ private information storage schemes [36] to solve this problem. As for hardware-based approaches, the secure coprocessor can receive location updates from the users subscribed to the service and update user locations during each database reshuffling [16]. However, both of these approaches demand additional computing resources from the already computationally constrained techniques.

8   Conclusion and Future Directions

The PIR-based approaches to location privacy presented in this chapter open door to a novel way of protecting user privacy. However, as mentioned in Section 7, full privacy guarantees of these approaches come at the cost of computationally intensive query processing. Therefore, reducing the costs associated with

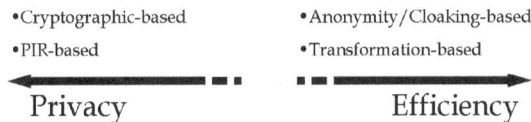

Fig. 12. The Privacy/Efficiency Spectrum

PIR operations can greatly increase the popularity of these approaches. For the approximate nearest neighbor queries discussed in Section 6.2, utilizing the excessive object information returned to a user to guarantee exact results is one promising research direction. As for the hardware-based approaches, employing more efficient shuffling techniques and moving as much non-secure processing as possible away from the SC can result in significant improvements.

Figure 12 summarizes the privacy/efficiency tradeoffs of various location privacy approaches discussed in this chapter. While anonymity/cloaking and transformation-based approaches enable efficient spatial query processing, they suffer from various privacy implications. At the other end of the spectrum, cryptographic and PIR-based approaches provide significantly stronger privacy guarantees by incurring more costly query processing operations. Therefore, developing a framework that strikes a compromise between these two extremes remains an interesting open problem. Such a framework should benefit from highly efficient spatial query processing while strongly protecting private user locations without any need for a trusted intermediary. Furthermore, expanding the current framework to efficiently support querying dynamic data remains a challenge. In addition to supporting dynamic queries, the aforementioned approaches can be generalized to support a wide range of spatial queries such as reverse nearest neighbor search and spatial joins.

References

1. Sweeney, L.: k-Anonymity: A Model for Protecting Privacy. Int. J. of Uncertainty, Fuzziness and Knowledge-Based Systems 10(5), 557–570 (2002)
2. Gruteser, M., Grunwald, D.: Anonymous usage of location-based services through spatial and temporal cloaking. In: MobiSys 2003, San Francisco, CA (2003)
3. Gruteser, M., Liu, X.: Protecting privacy in continuous location-tracking applications. IEEE Security & Privacy 2(2), 28–34 (2004)
4. Mokbel, M.F., Chow, C.Y., Aref, W.G.: The new casper: Query processing for location services without compromising privacy. In: VLDB 2006, Seoul, Korea, pp. 763–774 (2006)
5. Bettini, C., Wang, X.S., Jajodia, S.: Protecting privacy against location-based personal identification. In: Jonker, W., Petković, M. (eds.) SDM 2005. LNCS, vol. 3674, pp. 185–199. Springer, Heidelberg (2005)
6. Gedik, B., Liu, L.: A customizable k-anonymity model for protecting location privacy. In: ICDCS 2005, Columbus, OH, pp. 620–629 (2005)
7. Beresford, A.R., Stajano, F.: Location privacy in pervasive computing. IEEE Pervasive Computing 2(1), 46–55 (2003)

8. Khoshgozaran, A., Shahabi, C.: Blind evaluation of nearest neighbor queries using space transformation to preserve location privacy. In: Papadias, D., Zhang, D., Kollios, G. (eds.) SSTD 2007. LNCS, vol. 4605, pp. 239–257. Springer, Heidelberg (2007)

9. Ghinita, G., Kalnis, P., Khoshgozaran, A., Shahabi, C., Tan, K.L.: Private queries in location based services: anonymizers are not necessary. In: SIGMOD 2008, Vancouver, BC, Canada, pp. 121–132 (2008)

10. Khoshgozaran, A., Shirani-Mehr, H., Shahabi, C.: SPIRAL, a scalable private information retrieval approach to location privacy. In: The 2nd International Workshop on Privacy-Aware Location-based Mobile Services (PALMS) in conjunction with MDM 2008, Beijing, China (2008)

11. Hengartner, U.: Hiding location information from location-based services. In: MDM 2007, Mannheim, Germany, pp. 268–272 (2007)

12. Yiu, M.L., Jensen, C.S., Huang, X., Lu, H.: Spacetwist: Managing the trade-offs among location privacy, query performance, and query accuracy in mobile services. In: ICDE 2008, Cancún, México, pp. 366–375 (2008)

13. Zhong, S., Li, L., Liu, Y.G., Yang, Y.R.: Privacy-preserving location-based services for mobile users in wireless networks. Technical report, Yale Univerisity (2004)

14. Indyk, P., Woodruff, D.P.: Polylogarithmic private approximations and efficient matching. In: Halevi, S., Rabin, T. (eds.) TCC 2006. LNCS, vol. 3876, pp. 245–264. Springer, Heidelberg (2006)

15. Zhong, G., Goldberg, I., Hengartner, U.: Louis, lester and pierre: Three protocols for location privacy. In: Borisov, N., Golle, P. (eds.) PET 2007. LNCS, vol. 4776, pp. 62–76. Springer, Heidelberg (2007)

16. Khoshgozaran, A., Shahabi, C., Shirani-Mehr, H.: Location privacy; moving beyond k-anonymity, cloaking and anonymizers. Technical report, University of Southern California (2008)

17. Chor, B., Goldreich, O., Kushilevitz, E., Sudan, M.: Private information retrieval. In: FOCS, pp. 41–50 (1995)

18. Kushilevitz, E., Ostrovsky, R.: Replication is not needed: Single database, computationally-private information retrieval. In: FOCS, pp. 364–373 (1997)

19. Sion, R.: On the computational practicality of private information retrieval. In: Proceedings of the Network and Distributed Systems Security Symposium, 2007. Stony Brook Network Security and Applied Cryptography Lab. Tech. Report (2007)

20. Asonov, D.: Querying Databases Privately. LNCS, vol. 3128. Springer, Heidelberg (2004)

21. Asonov, D., Freytag, J.C.: Almost optimal private information retrieval. In: Dingledine, R., Syverson, P.F. (eds.) PET 2002. LNCS, vol. 2482, pp. 209–223. Springer, Heidelberg (2003)

22. Iliev, A., Smith, S.W.: Private information storage with logarithm-space secure hardware. In: International Information Security Workshops, Toulouse, France, pp. 199–214 (2004)

23. Smith, S.W., Safford, D.: Practical private information retrieval with secure coprocessors. Technical report, IBM (August 2000)

24. Gertner, Y., Goldwasser, S., Malkin, T.: A random server model for private information retrieval or how to achieve information theoretic pir avoiding database replication. In: Rolim, J.D.P., Serna, M., Luby, M. (eds.) RANDOM 1998. LNCS, vol. 1518, pp. 200–217. Springer, Heidelberg (1998)

25. Beimel, A., Ishai, Y., Malkin, T.: Reducing the servers' computation in private information retrieval: Pir with preprocessing. J. Cryptology 17(2), 125–151 (2004)

26. Gertner, Y., Ishai, Y., Kushilevitz, E., Malkin, T.: Protecting data privacy in private information retrieval schemes. J. Comput. Syst. Sci. 60(3), 592–629 (2000)
27. Bhattacharjee, B., Abe, N., Goldman, K., Zadrozny, B., Chillakuru, V.R., del Carpio, M., Apte, C.: Using secure coprocessors for privacy preserving collaborative data mining and analysis. In: DaMoN 2006, Chicago, IL, pp. 1–7 (2006)
28. Jiang, S., Smith, S., Minami, K.: Securing web servers against insider attack. In: ACSAC 2001, Washington, DC, USA, pp. 265–276 (2001)
29. Kalashnikov, D.V., Prabhakar, S., Hambrusch, S.E.: Main memory evaluation of monitoring queries over moving objects. Distrib. Parallel Databases 15(2), 117–135 (2004)
30. Xiong, X., Mokbel, M.F., Aref, W.G.: Sea-cnn: Scalable processing of continuous k-nearest neighbor queries in spatio-temporal databases. In: ICDE 2005, Tokyo, Japan, pp. 643–654 (2005)
31. Yu, X., Pu, K.Q., Koudas, N.: Monitoring k-nearest neighbor queries over moving objects. In: ICDE 2005, Tokyo, Japan, pp. 631–642 (2005)
32. Hilbert, D.: Uber die stetige abbildung einer linie auf ein flachenstuck. Math. Ann. 38, 459–460 (1891)
33. Faloutsos, C., Roseman, S.: Fractals for secondary key retrieval. In: PODS 1989: Proceedings of the eighth ACM SIGACT-SIGMOD-SIGART symposium on Principles of database systems, New York, NY, USA, pp. 247–252 (1989)
34. Flath, D.E.: Introduction to Number Theory. John Wiley & Sons, Chichester (1988)
35. Berg, M.d., Kreveld, M.v., Overmars, M., Schwarzkopf, O.: Computational geometry: Algorithms and applications. Springer, Heidelberg (1997)
36. Ostrovsky, R., Shoup, V.: Private information storage (extended abstract). In: STOC 1997, New York, NY, USA, pp. 294–303 (1997)

Privacy Preservation over Untrusted Mobile Networks

Claudio A. Ardagna[1], Sushil Jajodia[2], Pierangela Samarati[1],
and Angelos Stavrou[2]

[1] Dipartimento di Tecnologie dell'Informazione
Università degli Studi di Milano
Via Bramante, 65 - Crema, Italy
{claudio.ardagna,pierangela.samarati}@unimi.it
[2] CSIS - George Mason University
Fairfax, VA, USA 22030-4444
{jajodia,astavrou}@gmu.edu

Abstract. The proliferation of mobile devices has given rise to novel user-centric applications and services. In current mobile systems, users gain access to remote servers over mobile network operators. These operators are typically assumed to be trusted and to manage the information they collect in a privacy-preserving way. Such information, however, is extremely sensitive and coveted by many companies, which may use it to improve their business. In this context, safeguarding the users' privacy against the prying eyes of the network operators is an emerging requirement.

In this chapter, we first present a survey of existing state-of-the-art protection mechanisms and their challenges when deployed in the context of wired and wireless networks. Moreover, we illustrate recent and ongoing research that attempts to address different aspects of privacy in mobile applications. Furthermore, we present a new proposal to ensure private communication in the context of hybrid mobile networks, which integrate wired, wireless and cellular technologies. We conclude by outlining open problems and possible future research directions.

1 Introduction

Recent advancements in mobile sensing technologies and the growth of wireless and cellular networks have radically changed the working environment that people use to perform everyday tasks. Today, people are used to be online and stay connected independently of their physical location. This ubiquitous connectivity empowers them with access to a wealth of mobile services. Furthermore, the ease of use of mobile e-commerce and location-based services has fostered the development of enhanced mobile applications [1,2,3].

Unfortunately, the pervasiveness, the accuracy, and the broadcast nature of wireless technologies can easily become the next privacy attack vector, exposing a wide-range of information about everyday activities and personal lives

C. Bettini et al. (Eds.): Privacy in Location-Based Applications, LNCS 5599, pp. 84–105, 2009.
© Springer-Verlag Berlin Heidelberg 2009

to unauthorized eyes. The worst case scenario that analysts have foreseen as a consequence of an unrestricted and unregulated availability of wireless technologies recalls the "Big Brother" stereotype: a society where the secondary effect of wireless technologies – whose primary effect is to enable the development of innovative and valuable services – becomes a form of implicit total surveillance of individuals. Today, this "Big Brother" scenario is becoming more and more a reality rather than just a prediction. Some recent examples can provide an idea of the extent of the problem. In September 2007, Capla Kesting Fine Art announced the plan of building a cell tower, near Brooklyn NY, able to capture, monitor and rebroadcast wireless signals and communications to ensure public safety [4]. In addition, in 2007, the US Congress approved changes to the 1978 Foreign Intelligence Surveillance Act giving to NSA the authorization to monitor domestic phone conversations and e-mails including those stemming from the cellular network and the Internet. This legislation provides the legal grounds for the cell tower's construction and for the monitoring of users communications in the cellular network. Furthermore, there are numerous examples of rental companies that employed GPS technology to track cars and charge users for agreement infringements [5], or organizations using a location service to track their own employees [6]. The question of what constitutes a legitimate and user-approved use of the mobile tracking technology remains unclear and can only become worse in the near future.

In today's scenario, concerns about the protection of users' privacy represent one of the main reasons that limit the widespread diffusion of mobile services. Although the need of privacy solutions for mobile users arises, existing solutions are only palliative and weak in mobile contexts. Privacy solutions in fact primarily focus on protecting the users against services that collect the users' personal data for service provisioning. However, the advent of cellular (and in general hybrid) networks has made the problem of protecting the users' privacy worse: users should also be protected from the prying eyes of mobile peers and mobile network operators. The operators are in a privileged position, able to observe and analyze each communication on the network. As a consequence, they have the capability to generate, share, and maintain precise profiles of the users over long periods of time. Such profiles include personal information, such as, for instance, servers visited and points of interest, shopping and travel habits among other things. This scenario introduces a new set of requirements to be addressed in the protection of users' privacy. In particular, there is a pressing need for a mechanism that protects the communication privacy of mobile users. Such a mechanism should depart from the traditional privacy view, and consider a new threat model including operators and peers as potential adversaries. This new view of the problem is especially valid in the context of mobile hybrid networks, where users can communicate on different networks (e.g., wired, WiFi, cellular).

The remainder of this chapter is organized as follows. Section 2 illustrates basic concepts on network privacy protection. Section 3 presents recent proposals and ongoing work addressing different privacy issues in distributed and mobile networks and applications. Section 4 discusses emerging trends and a new vision

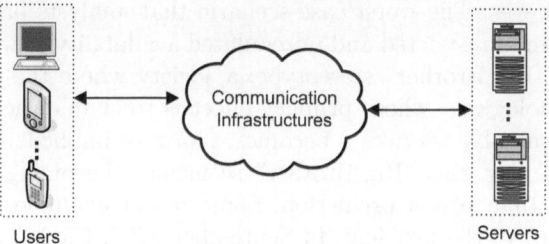

Fig. 1. Basic scenario

of privacy in the field of mobile hybrid networks, and presents a new approach for preserving communication privacy in hybrid networks. Section 5 presents open problems and future work. Finally, Section 6 concludes the chapter.

2   Basic Concepts on Network Privacy Protection

Regardless of the technology implemented, a network infrastructure is composed at an abstract level by three main entities (see Figure 1): *users*, who join the network to interact with and access, *servers* and *communication infrastructures*, that provide the platforms enabling communications between users and servers.

Research on distributed and mobile networks has traditionally focused on providing a communication infrastructure with high performance, efficiency, security, and reliability. Today, technology improvements provide solutions to efficiently store, mine, and share huge amount of users information, thus raising privacy concerns [7]. Privacy solutions are then needed and can be aimed at protecting different aspects of a communication, depending on the scenario and on the adversary model. In this chapter, we focus on protecting the information related to the fact that given parties communicate to each other (communication privacy). We do not discuss the problem of protecting the content of a communication (i.e., integrity and confidentiality), assuming that communication content can be protected by exploiting classical techniques [8]. Also, the vast amount of information exchanged, especially when users surf the Web, makes solutions that protect only communications content inadequate. The privacy of the identities of the participating parties has to be also preserved.

Different protection paradigms have been defined for preserving the privacy of the communications. Typically, they are based on the concept of *anonymity*. Anonymity states that an individual (i.e., the identity or personally identifiable information of an individual) should not be identifiable within an *anonymity set*, that is, a set of users. In the context of network communications, the following protection paradigms have been defined [9].

– *Sender anonymity.* It refers to the communication originator: the identity of the sender of a message must be hidden to external parties (including the receiver itself).

- *Receiver anonymity.* It refers to the communication destination: the identity of the receiver of a message must be hidden to external parties (not including the sender).
- *Communication anonymity.* It encompasses sender and receiver anonymity: the identity of both the sender and receiver of a message must be hidden from external parties. An external party only knows that a communication is in place. Communication anonymity also includes the concept of *unlinkability*, meaning that an observer might know that the sender and receiver are involved in some communications on the network, but does not know with whom each of them communicates.

Similar protection paradigms can be introduced based on the concept of k-anonymity, rather than anonymity. k-anonymity has been originally defined in the context of databases [10,11] and captures a traditional requirement followed by statistical agencies according to which the released data should be indistinguishably related to no less than a certain number k of respondents. Adapting this concept to the context of networks, we can consider the definition of sender, receiver, and communication k-anonymity.

When the above paradigms are used, an important aspect to consider is the adversary against which anonymity is to be guaranteed. Several solutions have been developed to protect the privacy of the communication against *i)* the *servers* providing services, *ii) external parties* which can observe the communication, and *iii) internal observers* that reside in the network of the target user. Some works have also assumed the entities responsible for the management of the communication infrastructure (i.e., network operators) as potential adversaries [12]. This latter scenario poses an entirely different set of requirements in the context of mobile hybrid networks, and requires therefore careful consideration and ad-hoc solutions (Section 4).

3 Overview of Related and Ongoing Research

While the deployment and management of mobile networks have been considered in earlier research in the area of mobile applications, approaches aimed at protecting the privacy of users have gained great relevance only in the last few years. Furthermore, research in the context of mobile networks has typically approached the privacy problem from the perspective of providing anonymous communications. In this section, we first provide a survey of the solutions that offer communication anonymity in the context of wired networks and their problems when applied to mobile networks (Section 3.1). We then discuss two different lines of research on anonymity in mobile networks. First, we discuss techniques inspired by the work on wired networks (Section 3.2). These solutions are aimed at providing communication anonymity by means of anonymous routing algorithms in the context of mobile ad-hoc networks. Second, we discuss techniques to be used in the context of location-based services (Section 3.3). These approaches focus on protecting the sender anonymity at the application layer against untrusted servers.

3.1 Communication Anonymity in Wired Networks

Chaum introduces a technique based on public key cryptography and the concept of "mix" to provide sender anonymity and communication untraceability [13]. The basic idea consists in forwarding each communication from sender to receiver through one or more mixes, which form a *mix network*. A mix is responsible for collecting a number of messages from different senders, shuffle them, and forward them to the next destination (possibly another mix node) in random order. The main purpose of each mix node is then to break the link between ingoing and outgoing messages, making the end-to-end communication untraceable and its tracking impervious for the adversaries. In addition, each mix node only knows the node from which a message is received and the one to which the message is to be sent. This makes mix networks strong against malicious mixes, unless all the mixes in a message path from sender to receiver are compromised and collude with the adversary. The return path is statically determined by the message sender and forwarded as a part of the message sent to the receiver. The receiver uses it to communicate back to the sender, thus preserving the users anonymity. As a result, Chaum's mix network provides a solution where adversaries are not able to follow an end-to-end communication.

Onion routing is a solution that exploits the notion of mix network to provide an anonymous communication infrastructure over the Internet [14,15]. Onion routing provides connections resistant to traffic analysis and eavesdropping, and is well suited for real-time and bi-directional communications. In onion routing, the sender creates the path of the connection through the onion routing network by means of an *onion proxy* that knows the network topology. The proxy produces an anonymous path to the destination, composed by several *onion routers*, and an *onion*, that is, a data structure composed by a layer of encryption for each router in the path, to be used in the sender-receiver communication. Once the path and the onion are established the message is sent through the anonymous connection. Each onion router receiving the message, peels off its layer of the onion, thus identifying the next hop, and sends the remaining part of the onion to the next router. Onion routers are connected by permanent socket connections. Similarly to mixes in mix networks, onion routers only know the previous and next hops of a communication. At the end, the message reaches the receiver in plain-text. Backward communications happen on the same anonymous path. This solution provides anonymity against internal and external adversaries (i.e., Internet routers and onion routers, respectively), since an adversary is able neither to infer the content of the message nor to link the sender to the receiver. The network only observes that a communication is taking place. Figure 2 shows an example of anonymous connection [16]. Black computer represents an onion router, while white one an onion proxy. Thick lines represent encrypted connections and thin ones a socket connection in clear. Different connections involving the same sender may require the establishment of different anonymous connections. At the end of a communication, the sender sends a destroy message. The path is then destroyed and each router deletes any information it knows about it. TOR is a second generation onion

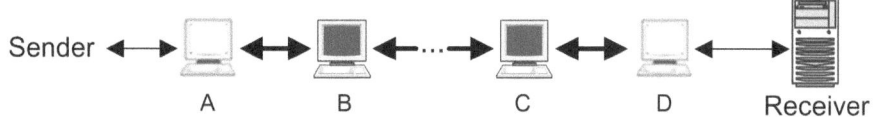

Fig. 2. Anonymous connection in an Onion Routing infrastructure

routing-based solution that provides anonymity by preventing adversaries from following packets from a sender to a receiver and vice versa [17]. In addition to traditional anonymous routing, TOR allows the sender to remain anonymous to the receiver. TOR addresses some limitations affecting the original design of onion routing by adding perfect forward secrecy, congestion control, directory servers, integrity checking, configurable exit policies, and a practical design for location-hidden services via rendezvous points [17]. In TOR, the onion proxy responsible to define the anonymous connection is installed on the user's machine. When the user needs to communicate with another party, the proxy establishes the anonymous path and generates the onion. Then, the message (including the onion) is sent through the path. Each router receiving the message removes, by using its private key, a layer of encryption to the onion to know its successor. At the end of the path, the receiver node retrieves the message in plain-text. Backward communications happen on the same anonymous connection.

Another anonymizing solution, designed for Web-communications, is Crowds [9]. In Crowds, the routing path and its length are dynamically generated. A user starts a process, called *jondo*, on her computer to join a *crowd* (i.e., a set of users) through a server, called *blender*. The blender receives a connection request from the jondo and decides if the jondo is allowed to join the crowd. If the jondo is admitted, it receives all the information to interact within the crowd. After this, the blender is no longer involved in the communication. All the user requests are sent to the jondo. The first request by a user is used to start the path establishment as follows. The user's jondo selects another jondo in the crowd (including itself) and forwards the request to it. Upon receiving the request, the receiving jondo either forwards the request to another jondo or sends it to the end server, with probability p_f. As a result, the request starts from the user's browser, follows a random number of jondos, and, eventually, is presented to the end server. As soon as a path is built, every request from the same jondo follows the same path, except for the end server (which may vary depending on to whom the user wants to send a message). The server response uses the same path as the user request. The path is changed when a new jondo joins the crowd or a jondo leaves it. Figure 3 shows a crowd composed of five jondos, on which two paths have been defined: 3→1→5→2→A (dotted lines) and 1→2→4→D (dashed lines). From an attacker point of view, the end server receiving a request cannot distinguish the sender among the users in the crowd. Also, collaborating users cannot know if a user is the sender or merely a node forwarding the request. Crowds is also robust against local eavesdroppers

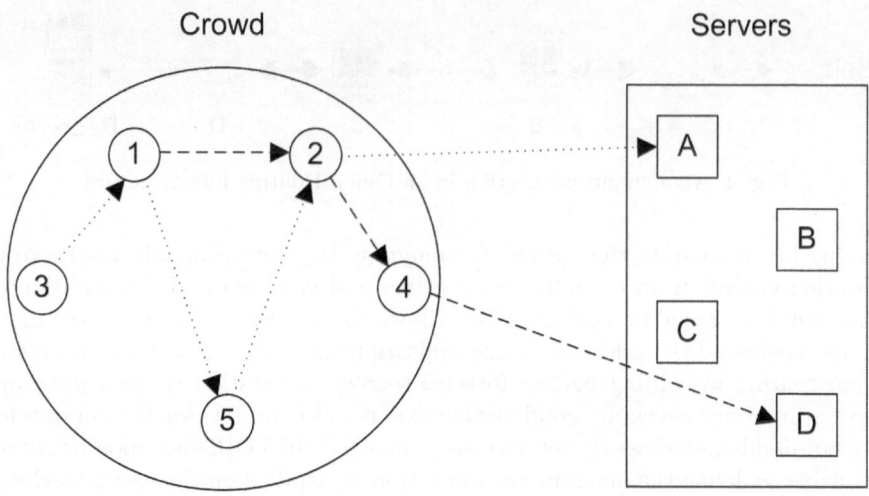

Fig. 3. Two paths in Crowds

that observe all the communications of a given node in a crowd. In fact, although a local eavesdropper can understand if a user is the sender, it never knows the receiver (i.e., the server), since the receiver resides in a different domain.

The solutions presented above aim at providing anonymous communications for protecting the privacy of the users in wired networks (e.g., Internet). Such solutions are not well suited for a mobile scenario, where users can wander freely while initiating transactions and communications by means of terminal devices like cell phones (GSM and 3G). In fact, solutions for wired networks: *1)* assume that the path generated by the sender is used both for the request and the response, *2)* assume a known network topology to create meaningful routes, and *3)* often rely on trusted third parties (e.g., mix, onion router, blender) and on heavy multiparty computation. These assumptions however do not hold for a mobile environment. In fact, mobile users: *1)* move fast over time, making the path used for the request likely to be not available both for the response, *2)* form networks of arbitrary topology, and *3)* use devices with limited capabilities, and then not suitable for solutions based on multiparty computation.

3.2   Communication Anonymity in Mobile Ad-Hoc Networks

In the context of mobile ad-hoc networks (MANETs), research on privacy protection has focused on preserving the privacy of wireless traffic by studying and providing privacy-enhanced and anonymous communication infrastructures. MANETs are composed by mobile routers and hosts that form networks of arbitrary topology, by means of wireless communications, and use ad-hoc routing protocols to communicate among them. The first routing protocols, such as AODV [18] and DSR [19], were not designed to provide or guarantee privacy and communication anonymity, rather they were aimed at increasing network performance, efficiency, security, and reliability. As a consequence, they are vulnerable to privacy violations, for instance, by exploiting the protocol state, since each node stores sender, receiver, and hop-count of each communication.

Subsequent work focused on routing protocols for mobile ad-hoc networks and attempted to protect anonymity and privacy. The solutions proposed did so by keeping secret to intermediate nodes the identities of the senders and receiver of messages. A number of anonymous routing protocols have then been presented [20,21,22,23,24,25,26]. Among them, MASK proposes an anonymous routing protocol, which provides both MAC-layer and network-layer communications without the need of using the real identities of the participating nodes [26]. MASK provides communication anonymity, in addition to node location anonymity and untraceability, and end-to-end flow untraceability. MASK relies on the use of dynamic pseudonyms rather than static MAC and network addresses, and on pairing-based cryptography to establish an anonymous neighborhood authentication between nodes and an anonymous network-layer communication. SDAR proposes a novel distributed routing protocol that guarantees security, anonymity and high reliability of the route [20]. SDAR relies on the encryption of packet headers and allows trustworthy intermediate nodes to participate in the path construction protocol without affecting the anonymity of the nodes involved in the communication. ANODR provides an untraceable and intrusion tolerant routing protocol [22]. It provides communication anonymity, by preventing adversaries from following packets in the network, and location privacy, by preventing the adversary to discover the real position of local transmitters (which could disclose also their identities). ANODR is based on the paradigm of "broadcast with trapdoor information". Discount-ANODR limits the overhead, suffered by ANODR, for providing sender anonymity and communication privacy [24]. A route is blindly generated by intermediary nodes, which only know the destination of the request and the identity of the immediately previous intermediary. Discount-ANODR provides a lightweight protocol based on symmetric key encryption and onion routing. No key exchange nor public key operations are needed. Capkun et al. propose a scheme for hybrid ad-hoc networks allowing users to communicate in a secure environment and preserve their privacy [27]. The authors assume privacy as composed of two parts: *i)* anonymity, which hides users identity in the network, and *ii)* location privacy, which protects the position of the users in the mobile environment. The solution proposed is based on continuously changing pseudonyms and cryptographic keys, it avoids users re-identification by observing the locations they visit, or the traffic they generate, and it provides secure and privacy-preserving communications in hybrid ad-hoc networks.

In the context of MANETs, a new type of ad hoc networks has been designed and developed, that is, Vehicular Ad-Hoc Networks (VANETs). VANETs, which are becoming more and more relevant and popular [28], consist of fixed equipments and vehicles equipped with sensors which form ad-hoc networks and exchange information, such as, for instance, traffic data and alarms. Traditional research in the context of VANET has ranged from the definition of efficient and reliable infrastructures to the development of enhanced applications.

Only recently, few works have focused on the security and privacy problems in VANETs [28,29,30,31]. Lack of security and privacy protection, in fact, can result in attacks subverting the normal network behaviour (e.g., by inserting false information) and violating the privacy of the users. Raya and Hubaux propose a preliminary investigation of the problem of guaranteeing security in VANET still protecting the privacy of the users [28]. They provide a threat model analyzing communication aspects, attacks, and security requirements. Also, they propose initial security solutions that protect user privacy based on digital signature, cryptographic keys, and anonymous public/private key pairs. Lin et al. present GSIS, a security and privacy solution based on Group Signature and Identity-based Signature techniques [30]. GSIS provides vehicle traceability to be used in case of disputes, and *conditional privacy preservation*. Conditional means that user-related information (e.g., driver's name, speed, position) must be accessible in case of exceptional situations, such as, crime or car accidents. Sampigethaya et al. present AMOEBA, a robust location privacy scheme for VANET [31]. AMOEBA focuses on protecting users privacy against malicious parties aiming at tracking vehicles, and building a profile of LBSs they access. To these aims, AMOEBA relies on vehicular groups and random silent periods.

The main limitation shared by the above solutions is that they heavily rely on key encryption, dynamic keys or pseudonyms, making them not always suitable in environments where communication devices have limited computational capabilities.

3.3 Sender Anonymity in Location-Based Services

Recent work on privacy protection has addressed the problem of preserving the anonymity of users (sender) that interact with Location-Based Services (LBSs) [32,33]. LBSs are considered untrusted parties that can exploit location information of users to breach their privacy. The main goal of most of the current solutions [34] is to guarantee anonymity, by preventing adversaries to use location information for re-identifying the users. In this scenario, each location measurement is manipulated to keep users' identity hidden, still preserving the best accuracy possible. The approaches discussed in the following are based on the notion of k-anonymity [10,11], which is aimed at making an individual not identifiable by releasing a geographical area containing at least k-1 users other than the requester. In this way, the LBSs cannot associate each request with fewer than k respondents, thus providing sender k-anonymity.

Bettini et al. propose a framework for evaluating the risk of disseminating sensitive location-based information, and introduce a technique aimed at supporting k-anonymity [35]. In this context, a *location-based quasi-identifier* (i.e., a set of attributes exploitable for linking) is defined as a set of spatio-temporal constraints, each one defining an area and a time window. The geo-localized history of the requests submitted by a user can be seen as a quasi-identifier, and used to discover sensitive information and re-identify the user. For instance, a user tracked during working days is likely to commute from her house to her workplace in a specific time frame in the morning, and to come back in another specific time frame in the evening. The notions of quasi-identifier and k-anonymity are used to provide a solution where a server collecting both the users' requests for services and the sequence of updates to users' locations, is not able to link a subset of requests to less than k users (sender k-anonymity). In other words, each data release must be such that every combination of values of quasi-identifiers can be indistinctly matched to at least k individuals. To this aim, there must exist k users having a personal history of locations consistent with the set of requests that has been issued.

Gruteser and Grunwald propose a middleware architecture and adaptive algorithms to comply with a given k-anonymity requirement, by manipulating location information, in spatial or temporal dimensions [36]. They consider a bi-dimensional space and introduce an algorithm based on quadtree partition method to decrease the spatial accuracy of location information (spatial cloaking). Spatial cloaking perturbs the location of the user by enlarging her real position. More in details, a middleware manages a geographical area including different users. When the location information of a requester needs to be manipulated for privacy protection, the middleware incrementally partitions the whole area on the x and y axis to achieve the requested k-anonymity with the best possible location accuracy, i.e., generating the smallest area containing k users (including the requester). In addition to spatial cloaking, a temporal cloaking algorithm perturbs the location information of the user in the temporal dimension. This algorithm produces more accurate spatial information, sacrificing the temporal accuracy. A further parameter, called spatial resolution, is defined to identify an area containing the requester. As soon as k-1 other users traverse this area, a time interval $[t_1, t_2]$ is generated and released with the area. By construction, in the interval $[t_1, t_2]$, k users, including the requester, have traversed the area identified by the spatial resolution parameter, thus satisfying preference k of the requester. Figure 4 shows an example of quadtree-based spatial cloaking. Let u_1 be a user with preference $k_1{=}3$ that submits a request. First, the spatial cloaking algorithm partitions the whole area in four quadrants (i.e., Q1, Q2, Q3, Q4). Second, the algorithm selects the quadrant containing u_1 (i.e., Q1), while it discards the others, and considers u_1's privacy preference. Since k_1 is enforced by Q1, Q1 is recursively partitioned in four quadrants (dashed line). This time, however, k_1 would not be satisfied and then Q1 is returned as the k-anonymous area. The same process is applied for user u_2 with preference $k_2{=}2$. In this case, the quadrant Q4.1 is retrieved as the anonymized user location. As a result, quadrant Q1 and Q4.1 provide sender k-anonymity.

Mokbel et al. present a framework, named Casper, which includes a *location anonymizer*, responsible for perturbing the location information of users to achieve k-sender anonymity, and a *privacy-aware query processor*, responsible for the management of anonymous queries and cloaked spatial areas [37]. In Casper, users define two parameters as privacy preferences: a degree of anonymity k, and the best accuracy A_{min} of the area that the user is willing to release. Two techniques which provide anonymization functionalities are implemented, that is, *basic* and *adaptive* location anonymizer. The main differences between the two

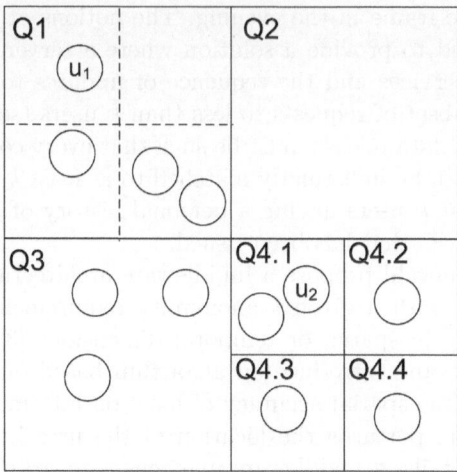

Fig. 4. Quadtree-based spatial cloaking

techniques lie in the data structures they use for anonymizing the users, and in their maintenance. The basic location anonymizer uses a pyramid structure. At each level of height h, 4^h cells are available; the root is at level $h=0$ and represents the whole area. Each cell has an identifier, and maintains track of the number of users within it. The system also maintains a hash table that stores information about users (identifiers, privacy profiles, and cell identifiers in which they are located). In the adaptive location anonymizer, the contents of the grid cells and of the hash table are the same. However, an incomplete pyramid data structure is maintained, with only the cells that can be potentially used as a cloaked area. Those cells for which no privacy preference needs to be enforced are not stored. Both the techniques implement a cloaking algorithm where the anonymized area is generated starting from the lowest level of the pyramid, and selecting the first cell that satisfies the preferences k and A_{min} of the sender.

Gedik and Liu describe a k-anonymity model and define a message perturbation engine responsible for providing location anonymization of user's requests through identity removal and spatio-temporal obfuscation of location information [38]. In this framework, each user defines a minimum level of anonymity to protect her privacy, and maximum temporal and spatial tolerances for preserving a level of quality of service. The message perturbation engine generates anonymous queries through the *CliqueCloak* algorithm. The CliqueCloak algorithm is based on a constraint graph where each vertex represents a message submitted by a user, and two vertices are connected if and only if the position of each user belongs to the constrained box of the other user, that is the area identified by the defined spatial tolerance. A valid k-anonymous perturbation of a message m is found if a set of at least other k-1 messages form an l-clique (i.e., a partition of the graph including l messages), such that the maximum k is less than l.

Ghinita et al. propose PRIVÈ, a decentralized architecture and an algorithm (*hilbASR*) for the protection of the sender anonymity of users querying LBSs [39].

The hilbASR algorithm is based on the definition of k-anonymous areas through the Hilbert space-filling curve. Specifically, 2D positions of users are mapped in 1D values, which are used to group users in buckets of k (anonymity areas). The hilbASR algorithm is strong against attackers who know the distribution of all users. This is achieved by satisfying the *reciprocity* property, which assures that if the hilbASR algorithm is applied to all users in an anonymity area, the same anonymity area is produced. PRIVÈ relies on a distributed B^+-tree with additional annotation to manage the definition of anonymized areas.

Hashem and Kulik present a decentralized approach to anonymity in a wireless ad-hoc network where each user is responsible for generating her cloaked area by communicating with others users [40]. The proposed approach combines k-anonymity with obfuscation. More in details, each peer: *1)* obfuscates her position by substituting the precise location with a locally cloaked area (LCA) and *2)* anonymizes her requests by manipulating the LCA to a global cloaked area (GCA). The GCA includes the LCAs of at least other k-1 users. An anonymous algorithm selects a query requester in the GCA with a near-uniform randomness, thus ensuring sender anonymity.

Cornelius et al. discuss the problem of protecting the privacy of the users involved in large-scale mobile applications that exploit collaborative and opportunistic sensing by mobile devices for service release [41]. In the proposed architecture, applications can distribute sensing works to anonymous mobile devices, and receive anonymized (but verifiable) sensor data in response.

Finally, Zhong and Hengartner present a distributed protocol for sender k-anonymity based on cryptographic mechanisms and secure multiparty computation [42]. The user interacts with multiple servers and a third party to determine if at least k people are in her area before communicating with the LBS. As a consequence, the LBS cannot re-identify the user. In addition, the servers involved in the anonymization process can infer neither the total number of users in the area nor if the k-anonymity property is satisfied (i.e., if at least k people including the user are in the area). Finally, the user can only know if the k-anonymity property holds.

Works on location k-anonymity share some limitations: *i)* they either rely on a centralized middleware for providing anonymity functionalities (centralized approach) or let the burden of the complexity in calculating the k-anonymous area to the users (decentralized approach); *ii)* they assume trusted mobile network operators; *iii)* they only provide k-anonymity at application level.

4   Privacy Protection in Mobile Hybrid Networks

In the previous section, we presented different approaches to protect the privacy of the users in different network scenarios, including wired networks, mobile ad-hoc networks, and mobile networks providing LBSs. In this section, we introduce an emerging scenario integrating all these network types, discuss a new adversary model where each party receiving part of the communication should be considered untrusted, and present a first solution to this privacy problem.

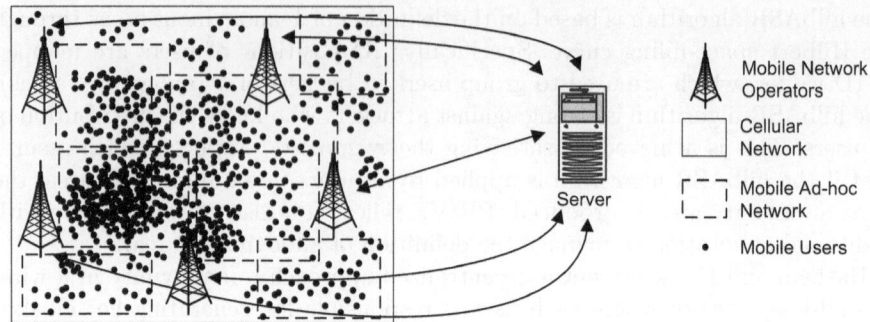

Fig. 5. Mobile network architecture

4.1   Basic Scenario

Already noted, previously proposed privacy protection systems mostly focused on protecting sender, receiver, or communication anonymity from untrusted servers and observers. They assume the network operators to be fully trusted. However, while it is reasonable to assume that the network operators are trusted with respect to the availability and working of the network, and to the management of communication data, since they have an incentive providing uninterrupted service, some trust cannot be put on the confidentiality of the data. In fact, personal users' information can be traded as a commodity and thus, network operators can no longer be trusted with safekeeping such information. This consideration is especially true for mobile hybrid networks [17,43,44] where a single infrastructure integrates heterogeneous technologies, such as, wireless, cellular, and wired technologies. Figure 5 shows the overall architecture that we take as a reference in the discussion. It includes the following participating entities.

- *Mobile Users.* Users carrying mobile devices supporting both GSM/3G and WiFi protocols for communication. They request services to servers available over the network.
- *Cellular Network* (and corresponding *Mobile Network Operators*). A network composed of multiple radio cells, which provide network access and services to mobile users. The cellular network acts as a gateway between mobile users and servers.
- *Servers.* Entities that provide online services to the mobile users and can collect their personal information for granting access.

Users can communicate via wireless and cellular protocols to access services that are either co-located in the cellular network or in the Internet. Mobile users establish ad-hoc (WiFi) point-to-point connections with other mobile peers in the network. As a result, there are several wireless Mobile Ad-Hoc Networks (MANETs), represented by the dashed rectangles in Figure 5. In addition, mobile users receive signals from the radio cells and can connect to the cellular networks, through which they can access services. Mobile users are registered with a given

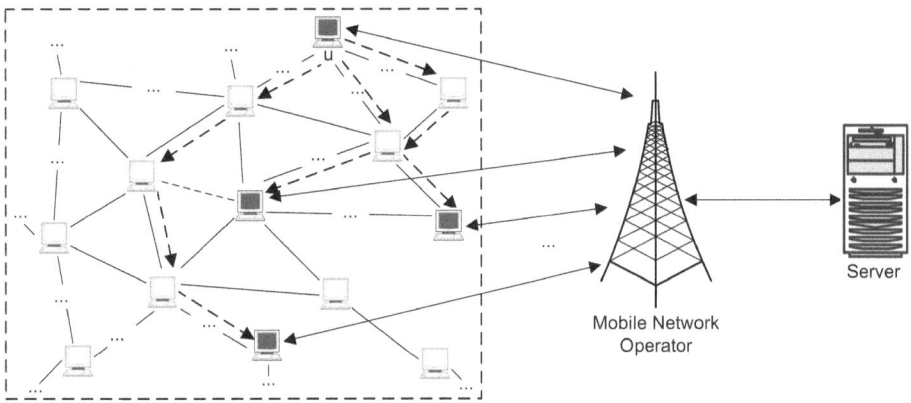

Fig. 6. A privacy-enhanced communication infrastructure

mobile network operator to access cellular functionality. Different users may use different mobile operators.

4.2   A New Vision of Privacy

A promising research direction for protecting privacy in mobile networks exploits the hybrid nature of current networks and the capabilities of mobile devices, which support both WiFi and cellular technologies, to provide anonymous communication protocols. In our proposal [12], we depart from the assumption of having a trusted mobile operator and exploit the intrinsic characteristics of hybrid networks to provide a privacy-enhanced communication infrastructure between users and servers (see Figure 6). All parties that can receive or observe communications, including the mobile operators through which users communicate with servers, are considered untrusted.[1] To address the privacy protection problem, we harness the fact that users can create WiFi point-to-point connections and at the same time join the cellular network in order to access the Internet through their mobile phones. Our solution is therefore different from the traditional research in anonymous communications [9,13,17,45], since is applicable to mobile hybrid infrastructure, and is aimed at protecting sender k-anonymity against mobile network operators.

4.3   A Multi-path Communication for Sender k-Anonymity

Our approach is based on k-anonymity and multi-path communication [12], to provide sender k-anonymity at network level. Sender k-anonymity is defined as follows.

Definition 1 (Sender k-anonymity). *Let M be a message originated by a mobile user u. User u is said to have sender k-anonymity, where k is the privacy preference of user u, if the probability of associating u as the message sender is less than or equal to $\frac{1}{k}$.*

---

[1] Depending on the scenario, each user can then decide if the server is trusted or not.

In the following, we show *i)* how a k-anonymous request is generated and transmitted by a mobile user to the server through mobile peers and the cellular network, thus exploiting a multi-path paradigm [46], and *ii)* how the server crafts a reply that can be received and decoded only by the requester concealed from the other k-1 peers, to protect sender k-anonymity against adversaries including mobile operators.

4.3.1   Overview of the Approach

Let $\mathcal{P}$, $\mathcal{O}$, and $\mathcal{S}$ be the set of mobile peers, mobile network operators, and servers in the hybrid network, respectively. In our discussion, user $u \in \mathcal{P}$ is the mobile peer that submits the request, $s \in \mathcal{S}$ the server, and $o \in \mathcal{O}$ the mobile network operator. Server s and the cellular network are in business relationship and u is subscribed to the cellular network. Also, s and u are assumed to be in a producer-consumer relationship and to share a common secret key SK that is generated through a Diffie-Hellman key exchange protocol. Each message M between u and s is encrypted, thus protecting confidentiality and integrity of the message through symmetric encryption (e.g., 3DES, AES). Standard notation $E_K()$ and $D_K()$ is used to denote encryption and decryption operations with key K. $E_{SK}(M)$ denotes a message M encrypted with symmetric key SK. Also, a random number mid is used as a message identifier. The complete protocol is shown in Figure 7 and is composed by an anonymous request and response, which are discussed in the following.

Anonymous Request. The anonymous request process is initiated by a mobile user u, which wishes to access a service provided by a server s. No overhead is given to u in the management of the mobile and anonymous process; u needs to first specify the message M and her privacy preference k. Then, u generates a message identifier mid and splits message M in k data flows producing a set of packets $\{m_1, m_2, \ldots, m_k\}$. The resulting packets are distributed among the neighbor mobile peers (peers for short) in the mobile ad-hoc network. Different algorithms (e.g., based on network state or on peer reputations) can be implemented for distributing packets among peers. Here, a simpler approach is used which consists in randomly forwarding the packets to the peers in u's communication range.

The distribution algorithm works as follows. Requester u encrypts each packet m_i using the symmetric key SK shared between u and s, and then appends mid in plain-text to it, that is, $\overline{m}_i = \{E_{SK}(m_i) \| mid\}$ for each $i=1\ldots k$. The presence of message id mid in every packet allows mobile peers to distinguish different packets belonging to the same message M. Requester u then randomly selects k-1 peers p in her communication range, and sends a packet (from $\overline{m}_2$ to $\overline{m}_k$) to each of them. It then sends $\overline{m}_1$ to s via o.

Upon receiving a packet $\overline{m}_i$ each peer p first checks mid. If she has already agreed to send a packet with the same mid (i.e., $mid \in$ SENT), p forwards $\overline{m}_i$ to another peer in the communication range. Otherwise, it randomly selects, with probability $p_f = \frac{1}{2}$, either to forward $\overline{m}_i$ to another peer in the communication range, or to send $\overline{m}_i$, without the mid, to s via o.

Initiator: Requester $u \in \mathcal{P}$
Involved Parties: Mobile peers $\mathcal{P}$, Mobile network operator o, Server s
Variables: Original message M, Response message M_r, Secret key SK shared between
$\qquad$ u and s

INITIATOR $(u \in \mathcal{P})$	u.1 Define message M and privacy preference k
	u.2 Generate a random number mid and split M in k packets $\{m_1, \ldots, m_k\}$
	u.3 Encrypt each packet m_i, with $i=1 \ldots k$, and append mid to them, $\overline{m}_i = [E_{SK}(m_i) \| mid]$
	u.4 **for** j:=2...k **do**
	$\qquad$ Select a peers $p_j \in \mathcal{P}$
	$\qquad$ Send to p_j a packet $\overline{m}_j$
	u.5 Select packet $\overline{m}_1$ and send it to o after a random delay
	u.6 Upon receiving response message M_r from o, decrypt M_r /*response*/
Peer $p \in \mathcal{P}$	p.1 Receive a packet $\overline{m}$
	p.2 **if** $\overline{m}.mid \in$ SENT
	$\qquad$ **then** forward $\overline{m}$ to a peer $p \in \mathcal{P}$
	$\qquad$ **else case** (p_f)
	$\qquad\qquad$ $\leq \frac{1}{2}$: forward $\overline{m}$ to a peer $p \in \mathcal{P}$
	$\qquad\qquad$ $> \frac{1}{2}$: $\overline{m} = \overline{m} - mid$
	$\qquad\qquad\qquad$ send $\overline{m}$ to o
	p.3 Upon receiving M_r from o, delete it /*response*/
Operator $o \in \mathcal{O}$	o.1 Receive a packet $\overline{m}$ from p
	o.2 Forward $\overline{m}$ to s
	o.3 Upon receiving M_r from s, forward it to p /*response*/
Server $s \in \mathcal{S}$	s.1 Receive a packet $\overline{m}$ from p via o
	s.2 Decrypt the packet with key SK and assemble M
	s.3 Generate and encrypt the response message M_r
	s.4 Send M_r to p through o /*response*/

Fig. 7. Anonymous communication protocol

After the distribution process, each selected peer p independently sends the packet received to s, through operator o. Operator o then sees packets that comes from k different peers, including u (who then remains k-anonymous), and forwards them to s. Now, server s can decrypt each packet, incrementally reconstruct the original message, and retrieve the user request.

Example 1. Figure 8(a) shows an example of communication. In the figure, white computer represents a peer that forwards a packet to another peer, while black one a peer that sends a packet to s. Requester u defines $k = 5$ and splits the message M in five parts $\{m_1, \ldots, m_5\}$. Packets are then encrypted with symmetric key SK shared between u and s, and mid is attached to each of them. Requester u sends packet $\overline{m}_1$ to s and forwards the other k-1 packets to peers in the communication range. Specifically, packets $\overline{m}_2$ and $\overline{m}_5$ are forwarded to peers p_1 and p_3 that send them to s. Assuming p_4 does not accept to send $\overline{m}_3$, packet $\overline{m}_3$ takes a forwarded path $p_4 \rightarrow p_7$. Packet $\overline{m}_4$ takes a forwarded path $p_6 \rightarrow p_7 \rightarrow p_9$ because, when the packet is received by p_7, p_7 notices that she has already accepted a packet ($\overline{m}_3$) with the same mid, and then forwards $\overline{m}_4$ to p_9. Finally, peers u, p_1, p_3, p_7, and p_9 send a packet to s via o.

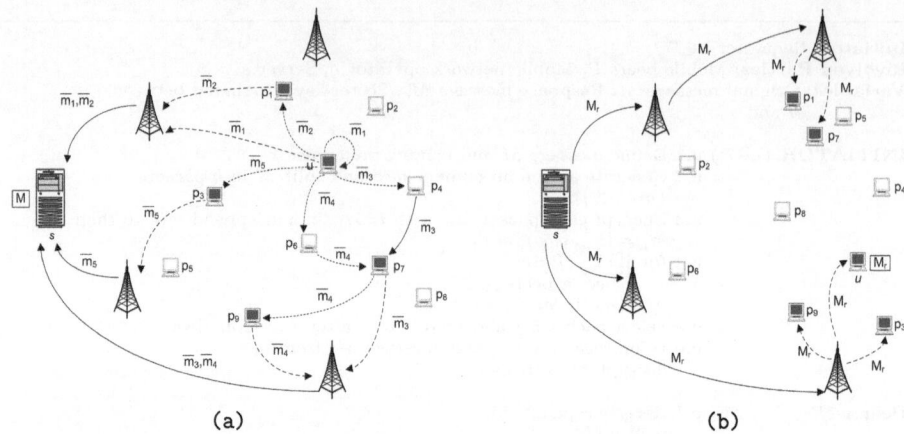

Fig. 8. Example of anonymous request (a) and anonymous response (b)

Anonymous Response. After the conclusion of the anonymous request process, server s retrieves the original message M and starts the service provisioning, which results in the release of an anonymous response to user u. The communication involves operator o to manage peers mobility and route the response to user u, still preserving her preference k. The anonymous response process works as follow. First of all, server s encrypts response message M_r with secret key SK shared with u. Then, it transmits the encrypted message M_r to the k peers involved in the anonymous request. Server s relies on the cellular network to manage the message delivery and the mobility of the peers. Although all peers receive the message, u is the only peer with secret key SK, and thus, she is the only one able to decrypt the message and benefit of the service.[2]

Example 2. Figure 8(b) shows an example of anonymous response to the request in Example 1. Encrypted message M_r is transmitted to all peers used in Example 1, that is, $\{u, p_1, p_3, p_7, p_9\}$. When u receives the message, she can decrypt it with key SK shared with the server. The other peers delete message M_r, since they are not able to open it.

The solution presented provides an *anonymous communication* protocol. In terms of anonymous communication, the message splitting and multi-path communication provide sender k-anonymity against mobile network operators. Also, the solution provides user accountability, since the user's identity is released to the server and can be retrieved by the operator when needed.

It is important to note that a privacy solution, to be practical, should not be invasive, requiring extensive modification of existing network protocols. Considering the solution described above, all the packets are routed regularly through

---

[2] To further strengthen the protocol, the server could potentially generate k-1 decoy messages, other than M_r. This can be performed by adding a *nonce* to the original message M_r before encrypting it with secret key SK. The cellular network sees k different response messages and it is not able to associate the response to the request.

the hybrid network using TCP and reconstructed at the destination server. Only some small changes are necessary and only for specific applications: the message splitting done by requester u, and the packet checks on the mobile ad-hoc network done by the peers.

5   Open Issues

We briefly describe some open problems which are important for the future development of privacy-enhanced and anonymous communication infrastructures for mobile networks.

- *Performance.* A key aspect for the success of privacy solutions in mobile networks is the performance and reliability of communications. The overhead in terms of end-to-end latency, the increase in the data transmission including both bursty and average bandwidth utilization should then be carefully evaluated. In addition, maintaining low power consumption is still an important performance metric for mobile and handheld devices with limited power. Finally, the performance evaluation should consider the adversarial and threat model and its impact on the performance metrics.
- *Malicious and uncooperative peers.* A complete and comprehensive privacy solution for mobile communications should consider malicious and uncooperative peers, which try to attack the system by modifying, dropping, injecting, or even replay received packets. An adversary model including malicious and uncooperative peers should then evaluate failure probability, that is, the probability to disrupt a communication given the rate of malicious peers in the environment surrounding the users. Finally, a complete model should evaluate the possibility of synchronized attacks, where malicious peers send a sequence of fake requests to neighbor peers trying to make their battery low.
- *Malicious mobile network operators.* The definition of untrusted mobile network operators is the most important paradigm shift with respect to traditional solutions developed for wired and mobile ad-hoc networks. An interesting research direction consists in exploring a solution which considers the possibility of malicious operators that modify, drop and replay received packets to expose communication anonymity and breach users' privacy.
- *Multiple rounds of communications.* An important aspect in the protection of the communication anonymity is the possibility of communications involving multiple rounds of request-response. In this case, intersection attacks can be used by an adversary to successfully expose the communication anonymity and link the user to a service request. Especially in the case of mobile networks, where users can move fast, randomly, and in a short time, intersection attacks become likely to be successful against anonymizing techniques. A strong solution should then provide countermeasures in case multiple rounds of request-response are needed for a service release.

- *Traffic accountability.* Traditionally, one main factor limiting the adoption of privacy solutions is the lack of a mechanism that makes the system accountable for the generated traffic and the operations at the server. In fact, servers are often reluctant to adopt privacy solutions that can be abused due to the lack of user accountability [47], or lack economic incentives. The problem is even worse when privacy solutions (e.g., anonymity techniques) completely hide the users. In addition to that, given the mobile scenario discussed in previous sections, a fundamental requirement is to provide the operators with the ability to distinguish genuine vs malicious traffic, detect malicious users, and keep them out of the network.
- *Participation in anonymizing networks.* An important aspect for the success of anonymizing networks is to foster users participation in them. A suitable solution should then provide automatic incentives, that is, the more a user collaborates in providing anonymity to other peers, the more protected is her communication.
- *Integration with anonymous services.* Solutions that provide communication anonymity against mobile network operators and mobile peers should maintain a level of integrability with existing solutions providing sender k-anonymity against the servers.
- *Multiparty computation.* In mobile networks, most of the existing privacy solutions and anonymous routing algorithms heavily rely on multiparty computation and cryptographic mechanisms. An important requirement for the success of these solutions consists in reducing the impact of multiparty computation on the end to end communication and on the power consumption.
- *Adversary knowledge.* A key aspect to be considered in the definition and development of a strong privacy solution in mobile networks is the effect of the adversary knowledge on the ability of an adversary to link a user to her services. For instance, personal information of users in an anonymity set can bring to situations in which the real requester is identified and associated to the service request.

6   Conclusions

In this chapter, we discussed and analyzed different aspects related to the protection of communication privacy for contemporary mobile networks. We discussed privacy issues in different applications and scenarios, focusing on: *i)* communication anonymity in wired and mobile networks; *ii)* preserving the privacy of wireless traffic through privacy-enhanced and anonymous routing protocols for MANET and VANET; and *iii)* protecting the privacy and anonymity of users that interact with untrusted LBSs. For all these areas, we presented the main solutions, and pointed out their peculiarities and open problems. Furthermore, in the context of mobile hybrid networks, we identified a promising research direction and a novel privacy-preserving scheme based on k-anonymity and multi-path communication, which aims at preserving privacy of users against mobile network operators. Finally, we brought forward some open problems that warrant further investigation.

Acknowledgments

This work was supported in part by the EU, within the 7th Framework Programme (FP7/2007-2013) under grant agreement no. 216483 "PrimeLife".

References

1. Barkhuus, L., Dey, A.: Location-based services for mobile telephony: a study of user's privacy concerns. In: Proc. of the 9th IFIP TC13 International Conference on Human-Computer Interaction (INTERACT 2003), Zurich, Switzerland (September 2003)
2. D'Roza, T., Bilchev, G.: An overview of location-based services. BT Technology Journal 21(1), 20–27 (2003)
3. Loopt (December 2008), http://www.loopt.com/about/privacy-security
4. Zander, C.: Cia Cell Tower Monitors Local Internet Users' Wireless Transmissions (September 2007), http://www.send2press.com/newswire/2007-09-0911-003.shtml
5. Chicago Tribune: Rental firm uses GPS in speeding fine, p. 9. Associated Press, Chicago (July 2, 2001)
6. Lee, J.W.: Location-tracing sparks privacy concerns. Korea Times, (November 16, 2004), http://news.naver.com/main/ read.nhn?mode=LPOD&mid=etc&oid=040&aid=0000016873
7. Giannotti, F., Pedreschi, D. (eds.): Mobility, data mining and privacy - Geographic knowledge discovery. Springer, Heidelberg (2008)
8. Kaufman, C., Perlman, R., Speciner, M.: Network security: Private communication in a public world. Prentice Hall, Englewood Cliffs (2003)
9. Reiter, M., Rubin, A.: Crowds: Anonymity for web transactions. ACM Transactions on Information and System Security 1(1), 66–92 (1998)
10. Ciriani, V., De Capitani di Vimercati, S., Foresti, S., Samarati, P.: k-Anonymity. In: Yu, T., Jajodia, S. (eds.) Secure Data Management in Decentralized Systems, Springer, Heidelberg (2007)
11. Samarati, P.: Protecting respondents' identities in microdata release. IEEE Transactions on Knowledge and Data Engineering 13(6), 1010–1027 (2001)
12. Ardagna, C., Stavrou, A., Jajodia, S., Samarati, P., Martin, R.: A multi-path approach for k-anonymity in mobile hybrid networks. In: Proc. of the International Workshop on Privacy in Location-Based Applications (PILBA 2008), Malaga, Spain (October 2008)
13. Chaum, D.: Untraceable electronic mail, return addresses, and digital pseudonyms. Communications of the ACM 24(2), 84–88 (1981)
14. Onion Routing, http://www.onion-router.net/
15. Reed, M., Syverson, P., Goldschlag, D.: Anonymous connections and onion routing. IEEE Journal on Selected Areas in Communications 16(4), 482–494 (1998)
16. Reed, M., Syverson, P., Goldschlag, D.: Proxies for anonymous routing. In: Proc. of the 12th Annual Computer Security Applications Conference, San Diego, CA (December 1996)
17. Dingledine, R., Mathewson, N., Syverson, P.: Tor: The Second-Generation Onion Router. In: Proc. of the 13th USENIX Security Symposium (August 2004)

18. Perkins, C., Royer, E.: Ad-hoc on demand distance vector routing. In: Proc. of the 2nd IEEE Workshop on Mobile Computing Systems and Applications (WMCSA 1999), New Orleans, LA, USA (February 1999)
19. Johnson, D.B., Maltz, D.A.: Dynamic Source Routing in Ad Hoc Wireless Networks, vol. 353. Kluwer Academic Publishers, Dordrecht (1996)
20. Boukerche, A., El-Khatib, K., Xu, L., Korba, L.: SDAR: A secure distributed anonymous routing protocol for wireless andmobile ad hoc networks. In: Proc. of the 29th Annual IEEE International Conference on Local Computer Networks (LCN 2004), Tampa, FL, USA (October 2004)
21. Kao, J.C., Marculescu, R.: Real-time anonymous routing for mobile ad hoc networks. In: Proc. of the Wireless Communications and Networking Conference (WCNC 2007), Hong Hong (March 2007)
22. Kong, J., Hong, X.: ANODR: Anonymous on demand routing with untraceable routes for mobile ad-hoc networks. In: Proc. of the 4th ACM International Symposium on Mobile Ad Hoc Networking and Computing (MOBIHOC 2003), Annapolis, MD, USA (June 2003)
23. Wu, X., Bhargava, B.: AO2P: Ad hoc on-demand position-based private routing protocol. IEEE Transaction on Mobile Computing 4(4) (July-August 2005)
24. Yang, L., Jakobsson, M., Wetzel, S.: Discount anonymous on demand routing for mobile ad hoc networks. In: Proc. of the Second International Conference on Security and Privacy in Communication Networks (SECURECOMM 2006), Baltimore, MD, USA (August-September 2006)
25. Zhang, Y., Liu, W., Lou, W.: Anonymous communication in mobile ad hoc networks. In: Proc. of the 24th Annual Joint Conference of the IEEE Communication Society (INFOCOM 2005), Miami, FL, USA (March 2005)
26. Zhang, Y., Liu, W., Lou, W., Fang, Y.: Mask: Anonymous on-demand routing in mobile ad hoc networks. IEEE Transaction on Wireless Communications 5(9) (September 2006)
27. Capkun, S., Hubaux, J.P., Jakobsson, M.: Secure and privacy-preserving communication in hybrid ad hoc networks. Technical Report IC/2004/10, EPFL-IC, CH-1015 Lausanne, Switzerland (January 2004)
28. Raya, M., Hubaux, J.P.: The security of vehicular ad hoc networks. In: Proc. of the 3rd ACM Workshop on Security of Ad hoc and Sensor Networks (SASN 2005), Alexandria, VA, USA (November 2005)
29. Dotzer, F.: Privacy issues in vehicular ad hoc networks. In: Danezis, G., Martin, D. (eds.) PET 2005. LNCS, vol. 3856, pp. 197–209. Springer, Heidelberg (2006)
30. Lin, X., Sun, X., Ho, P.H., Shen, X.: GSIS: A secure and privacy preserving protocol for vehicular communications. IEEE Transactions on Vehicular Technology 56(6), 3442–3456 (2007)
31. Sampigethaya, K., Li, M., Huang, L., Poovendran, R.: AMOEBA: Robust location privacy scheme for VANET. IEEE Journal on Selected Areas in Communications 25(8), 1569–1589 (2007)
32. Ardagna, C., Cremonini, M., Damiani, E., De Capitani di Vimercati, S., Samarati, P.: Supporting location-based conditions in access control policies. In: Proc. of the ACM Symposium on Information, Computer and Communications Security (ASIACCS 2006), Taipei, Taiwan (March 2006)
33. Ardagna, C., Cremonini, M., Damiani, E., De Capitani di Vimercati, S., Samarati, P.: Location privacy protection through obfuscation-based techniques. In: Barker, S., Ahn, G.-J. (eds.) Data and Applications Security 2007. LNCS, vol. 4602, pp. 47–60. Springer, Heidelberg (2007)

34. Mascetti, S., Bettini, C.: A comparison of spatial generalization algorithms for LBS privacy preservation. In: Proc. of the 1st International Workshop on Privacy-Aware Location-based Mobile Services (PALMS 2007), Mannheim, Germany (May 2007)

35. Bettini, C., Wang, X., Jajodia, S.: Protecting privacy against location-based personal identification. In: Jonker, W., Petković, M. (eds.) SDM 2005. LNCS, vol. 3674, pp. 185–199. Springer, Heidelberg (2005)

36. Gruteser, M., Grunwald, D.: Anonymous usage of location-based services through spatial and temporal cloaking. In: Proc. of the 1st International Conference on Mobile Systems, Applications, and Services (MobiSys 2003), San Francisco, CA, USA (May 2003)

37. Mokbel, M., Chow, C.Y., Aref, W.: The new casper: Query processing for location services without compromising privacy. In: Proc. of the 32nd International Conference on Very Large Data Bases (VLDB 2006), Seoul, Korea (September 2006)

38. Gedik, B., Liu, L.: Protecting location privacy with personalized k-anonymity: Architecture and algorithms. IEEE Transactions on Mobile Computing 7(1), 1–18 (2008)

39. Ghinita, G., Kalnis, P., Skiadopoulos, S.: PRIVÈ: Anonymous location-based queries in distributed mobile systems. In: Proc. of the International World Wide Web Conference (WWW 2007), Banff, Canada (May 2007)

40. Hashem, T., Kulik, L.: Safeguarding location privacy in wireless ad-hoc networks. In: Krumm, J., Abowd, G.D., Seneviratne, A., Strang, T. (eds.) UbiComp 2007. LNCS, vol. 4717, pp. 372–390. Springer, Heidelberg (2007)

41. Cornelius, C., Kapadia, A., Kotz, D., Peebles, D., Shin, M., Triandopoulos, N.: Anonysense: privacy-aware people-centric sensing. In: Proc. of the 6th international conference on Mobile systems, applications, and services (MobiSys 2008), Breckenridge, CO, USA (June 2008)

42. Zhong, G., Hengartner, U.: A distributed k-anonymity protocol for location privacy. In: Proc. of the Seventh IEEE International Conference on Pervasive Computing and Communication (PerCom 2009), Galveston, TX, USA (March 2009)

43. Fujiwara, T., Watanabe, T.: An ad hoc networking scheme in hybrid networks for emergency communications. Ad Hoc Networks 3(5), 607–620 (2005)

44. Sphinx - A Hybrid Network Model for Next Generation Wireless Systems, http://www.ece.gatech.edu/research/GNAN/work/sphinx/sphinx.html

45. Chaum, D.: The dining cryptographers problem: Unconditional sender and recipient untraceability. Journal of Cryptology 1(1), 65–75 (1988)

46. Stavrou, A., Keromytis, A.: Countering dos attacks with stateless multipath overlays. In: Proc. of the 12th ACM conference on Computer and communications security (CCS 2005), Alexandria, VA, USA (November 2005)

47. Borisov, N., Danezis, G., Mittal, P., Tabriz, P.: Denial of service or denial of security? In: Proc. of the 14th ACM conference on Computer and communications security (CCS 2007), Alexandria, Virginia, USA (October-November 2007)

Access Control in Location-Based Services

Claudio A. Ardagna, Marco Cremonini, Sabrina De Capitani di Vimercati,
and Pierangela Samarati

Dipartimento di Tecnologie dell'Informazione
Università degli Studi di Milano
Via Bramante, 65 - Crema, Italy
{claudio.ardagna,marco.cremonini,sabrina.decapitani,
pierangela.samarati}@unimi.it

Abstract. Recent enhancements in location technologies reliability and precision are fostering the development of a new wave of applications that make use of the location information of users. Such applications introduces new aspects of access control which should be addressed. On the one side, precise location information may play an important role and can be used to develop Location-based Access Control (LBAC) systems that integrate traditional access control mechanisms with conditions based on the physical position of users. On the other side, location information of users can be considered sensitive and access control solutions should be developed to protect it against unauthorized accesses and disclosures. In this chapter, we address these two aspects related to the use and protection of location information, discussing existing solutions, open issues, and some research directions.

1 Introduction

In the last decade, the diffusion and reliability achieved by mobile technologies have revolutionized the way users interact with the external world. Today, most people always carry a mobile device and can stay online and connected from everywhere. Location information is then available as a new class of users' information that can be exploited to develop innovative and valuable services (e.g., customer-oriented applications, social networks, and monitoring services). Several commercial and enterprise-oriented location-based services are already available and have gained popularity [1]. These services can be partitioned in different categories [2]. For instance, there are services that provide information on the position of the users or on the environment surrounding the location of a user (e.g., point of interest, traffic alerts), or services which can help in protecting human lives or highly sensitive information/resources. As an example, the enhanced 911 in North America [3] can exploit location information of users to immediately dispatch emergency services (e.g., emergency medical services, police, or firefighters) where they are needed, reducing the margin of error. In an environment offering location-based services (LBSs), *users* send a request for using such services to a *LBS provider*. The provider collects the user personal information, possibly interacting with a *location server* (LS), to decide whether the service can be granted

C. Bettini et al. (Eds.): Privacy in Location-Based Applications, LNCS 5599, pp. 106–126, 2009.

and how it can be possibly personalized. The location server works as the positioning system that measures the location information of users carrying mobile devices, and provides such information at different levels of granularity and with different Quality of Service (QoS). The types of location requests that a Location Server can satisfy depend on the specific mobile technology, the methods applied for measuring users position, and the environmental conditions.

Among the different issues that need to be addressed in the development of location-based services, *access control* is becoming increasingly important. Access control represents a key aspect to the success of location-based services, and can be radically changed by the availability of location information, which includes position and mobility of the users. In this chapter, access control issues are analyzed from two different perspectives: *1)* we analyze how current access control systems can integrate and exploit location information in evaluating and enforcing access requests, thus introducing Location-Based Access Control (LBAC) systems; *2)* we analyze how access control mechanisms should change for evaluating and enforcing access to location information, which might be *highly sensitive.*

In the first case, precise and accurate location information is used to enhance and strengthen access control systems by adding functionalities for defining, evaluating, and enforcing location-based policies, i.e., access control restrictions based on the position of the users. LBAC extends access control to the consideration of contextual location information, in particular the location of the user requesting access. Obtaining reliable and accurate location information with software applications reachable via a telecommunication infrastructure (e.g., wireless network) is a challenging aspect due to the intrinsic error of location measurements. An important requirement is then to provide a way to perform *location verification*, meaning that the location of a user has to be securely verified to meet certain criteria (e.g., being inside a specific room or within a geographical area). A stable and reliable verification mechanism can represent an important driver towards the development of a location-based access control system. Once a user's location has been verified using a protocol for location verification, the user can be granted access to a particular resource according to the desired policy. The location verification process must be able to tolerate rapid context changes, since mobile users, involved in transactions by means of their mobile devices, can wander freely and change their position in the network.

In the second case, location-based information is considered sensitive and therefore needs to be protected against unregulated access and disclosure. The unauthorized release of location information can result in several privacy breaches (e.g., [4]), and make the users target of fraudulent attacks [5] such as *unsolicited advertising*, when products and services are advertised by exploiting the user position without her consent; *physical attacks or harassment*, when the location of a user is used to carry physical assaults; and *users profiling*, when the location of a user is used to infer other sensitive information. This scenario poses a new set of requirements that need to be accomplished by access control systems for protecting location information. For instance, access control may be enriched with mechanisms that obfuscate the location information before its release to other parties [6,7]. Also, access control systems should be able to manage time-variant information, since location of users can change over time.

The remainder of this chapter is organized as follows. Section 2 describes the basic concepts of access control languages. Section 3 introduces the concept of location-based access control and describes some solutions implementing LBAC. Section 4 provides an overview of existing approaches to protect and manage access and disclosure of location information. Section 5 presents open problems and future work. Finally, Section 6 gives our conclusions.

2   Access Control Languages

Access control systems are based on *policies* that define authorizations concerning access to data/services. Authorizations establish who can (positive authorizations), or cannot (negative authorizations), execute which actions on which resources [8]. Recent advancements allow the specifications of policies with reference to generic attributes/properties of the parties (e.g., name, citizenship, occupation) and the resources (e.g., owner, creation date) involved [9,10,11]. A common assumption is that these properties characterizing users and resources are stored in *profiles* that define the name and the value of the properties. Users may also support requests for certified data (i.e., credentials), issued and signed by authorities trusted for making statements on the properties, and uncertified data, signed by the owner itself. For instance, an authorization can state that "a user of age greater than 18 and with a valid credit card number (requester) can read (action) a specific set of data (resource)". When an access request is submitted to the access control system, it is evaluated against the authorizations applicable to it.

From a modeling point of view, each authorization can be seen as a triple of the form ⟨*subject, object, actions*⟩, whose elements are generic boolean formulas over the subject requesting access, the object to which access is requested, and the actions the subject wants to perform on the object. The *subject* is an expression that allows referring to a set of subjects satisfying certain conditions, where conditions can evaluate the user's profile/properties, or the user's membership in groups, active roles, and so on. The *object* is an expression that allows referring to a set of objects satisfying certain conditions, where conditions evaluate membership of the object in categories, values of properties on metadata, and so on. The conditions specified in the policies can be built over generic predicates that can evaluate the information stored at the site or can evaluate state-based information (e.g., the role adopted inside an application, the number of access to a given object, time/date restrictions). For instance, an authorization stating that "professors with age greater than 35 can read critical documents created before the 2008" can be expressed as:

- subject: `equal(job,`*Professor*`)` ∧ `greater_than(age,`*35*`)`
- object: `equal(level,`*critical*`)` ∧ `less_than(creation,`*2008/01/01*`)`
- actions: *read*

where we assume that `equal`, `greater_than`, `less_than` are pre-defined predicates used to evaluate information stored in the user and/or object profiles, and whose semantic is self-explanatory. Access control policies can then be implemented by using different languages, like logic-based languages (e.g., [12]), which are expressive and characterized by a formal foundation, or XML-based languages (e.g., [9,11]), which are more suited to the Internet context.

In the next section, we discuss how access control policies based on boolean formula of conditions can be enriched by adding *location-based conditions*, which are expressed using ad-hoc location predicates. In the discussion, we do not make any assumption about the specific language used for implementing the policies and we refer to the abstract model just described.

3   Location-Based Access Control Systems

The diffusion and reliability reached by mobile technologies provide a means to use location information for improving access control systems in a novel way. Although, research on LBAC is a recent topic, the notion of LBAC is in itself not new. Some early mobile networking protocols already relied on linking the physical position of a terminal device with its capability of accessing network resources [13]. Extensive adoption of wireless local networks has triggered new interests in this topic. Some studies focused on location-based information for monitoring users movements on Wireless Lan [14] and 802.11 Networks [15]. Myllymaki and Edlund [16] describe a methodology for aggregating location data from multiple sources to improve location tracking features. Other researchers have investigated a line closer to LBAC by describing the architecture and operation of an access server module for access control in wireless local networks [1,17,18]. Controlling access to wireless networks, complying with IEEE 802.11 family protocols, is principally aimed at strengthening the well-known security weaknesses of wireless network protocol rather than at defining a general, protocol-independent model for LBAC. The need for a protocol-independent location technique has been highlighted by a study exploiting heterogeneous positioning sources like GPS, Bluetooth, and WaveLAN for designing location-aware applications [18]. Cho et al. [17] present a location-based protocol (Location-Based network Access Control) for authentication and authorization, in infrastructure-based WLAN systems based on IEEE 802.11. The protocol is used to securely authenticate the location claims released by wireless users, and exchange the keys shared for data encryption. The infrastructure is composed of three parties: the *key server* responsible for authentication, location claim verification, and key distribution, the *access points*, and the *mobile stations*. The solution is based on the fact that a mobile station is in a given location if and only if it receives all the relevant information from the corresponding access points. The protocol uses a Diffie-Helmann algorithm to authenticate location claims, authorize network access, and generate the shared keys for communications between mobile stations and access points. Location-based information and its management have been also the subject of a study by Varshney [1] in the area of mobile commerce applications. This is a related research area that has strong connection with location systems and is a promising source of requirements for LBAC models.

Other papers consider location information as a means for improving security. Sastry et al. [19] exploit location-based access control in sensor networks. Zhang and Parashar [20] propose a location-aware extension to Role-Based Access Control (RBAC) suitable for grid-based distributed applications. Atallah et al. [21] study the problem of key management and derivation in the context of geospatial access control. In this work, a geographical space is modeled as a grid of $m \times n$ cells and policies are used to define whether users can access a given rectangular spatial area composed of one or more cells. Each cell is associated with a key and contains information of interests for the users. When a user gains access to an area, a set of keys is derived. Each key enables the user to access a different cell in the area together with its information. Here, a user location is treated as a single point without explicitly considering the intrinsic uncertainty of location measurements. Atluri et al. [22] consider the problem of providing an efficient security policy enforcement for mobile environments. The authors briefly introduce an authorization model based on moving entities and spatio-temporal attributes, and consider three types of authorizations: *i)* on moving subjects and static objects, *ii)* on static subjects and moving objects, and *iii)* on moving subjects and moving objects. The paper concentrates on the enforcement of such authorizations by providing data structures suitable for the management of moving entities, and spatio-temporal authorizations. The paper presents an index structure called S^{PPF} that maintains past, present, and future locations of moving entities together with authorizations, using a partial persistent storage. An evaluation approach is then described where authorizations are compared with nodes modeling moving entities, by analyzing the spatio-temporal extents of both authorizations and moving entities. This solution allows efficient evaluation of access requests that also include *locate* and *track* privileges.

While all these approaches have made significant steps in the development of models and systems supporting location-based information, the definition of a LBAC model that takes into consideration the special nature of location information is still an emerging research issue that has not been yet fully addressed by the security and access control research community. Only few works provides solutions for defining and evaluating location-based policies. In the following, we first describe a solution providing a LBAC infrastructure [23] (Section 3.1) and then an extension to XACML [11] for the definition of geospatial predicates [24] (Section 3.2).

3.1   An Access Control System for LBAC Policies

Ardagna et al. [23] define a LBAC system that supports location-based policies. Intuitively, a location-based policy exploits the physical location of users to define when they can access a service or a resource. The authors identify three main steps towards the development of a LBAC system: *i)* the design of a reference LBAC architecture that can support the evaluation and enforcement of location-based policies; *ii)* the definition of location-based conditions; and

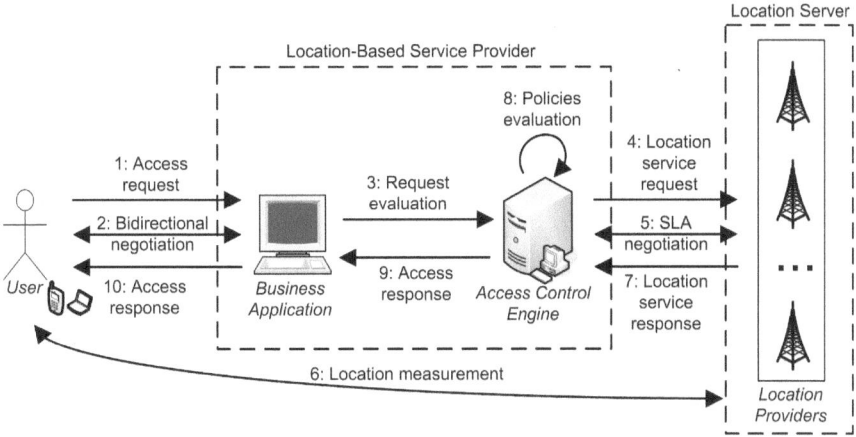

Fig. 1. LBAC architecture

iii) the definition of a mechanism for the evaluation and enforcement of location-based conditions.

3.1.1 LBAC Architecture

LBAC definition changes the conventional access control architecture, since there are more parties involved. Figure 1 presents a LBAC architecture that involves four logical components.

User. The entity whose access request to a location server must be authorized by a LBAC system. Users carry terminals enabling authentication and some form of location verification.

Business application. Customer-oriented application that offers services whose release is regulated by location-based policies.

Access Control Engine (ACE). The entity that is responsible for evaluating access requests according to some location-based policies. The ACE communicates with one or more Location Providers for acquiring location information. The ACE does not have direct access to the location information; rather, it sends requests to external services and waits for the corresponding answers.

Location Providers (LPs). The trusted entities that provide the location information (e.g., context data about location and timing, location-based predicate evaluation) by implementing Location Server interfaces.

Interactions among the User, the Business Application, the Access Control Engine, and the Location Providers are carried out via request/response messages (see Figure 1). The process is initiated by a user that submits an access request to a Business Application (step 1). A negotiation process between the two parties is then used to exchange those data that are relevant to the policy evaluation (step 2). The request is further forwarded to the ACE (step 3) that interacts (if needed) with the Location Providers (steps 4-7), evaluates policies (step 8), and returns an access decision (steps 9-10). Communications between the ACE and the Location Providers may be driven by a service level agreement (SLA) negotiation phase (step 5). This negotiation is used to agree upon and set quality of services attributes and the corresponding service cost.

3.1.2 Location-Based Conditions

The location-based conditions that might be useful to include in access control policies and whose evaluation is possible with today's technologies fall within three main categories:

- *position-based* conditions on the location of the users (e.g., to evaluate whether users are in a certain building or city, or in the proximity of other entities);
- *movement-based* conditions on the mobility of the users (e.g., velocity, acceleration, or direction where users are headed);
- *interaction-based* conditions involving relationships among multiple users or entities (e.g., the number of users within a given area).

Table 1 presents some specific predicates corresponding to the conditions of the kind identified by the classes above. In particular, predicates inarea, disjoint, and distance are of type position and evaluate the location of the users; velocity is of type movement and evaluates the mobility of the users; density and local_density are of type interaction and evaluate spatial relationships between entities. Other predicates can be added as the need arises and technology progresses. Conditions are expressed as boolean queries of the form *predicate(parameters,value)*, stating whether *predicate* over *parameters* has the specified *value*. The evaluation of a boolean query returns a triple *[bool_value,confidence,timeout]* stating whether the predicate is true or false (*bool_value*), the time validity associated with the assessment (*timeout*), and a *confidence* value expressing the reliability associated with the assessment. This confidence may depend on different aspects such as the accuracy, environmental and weather conditions, granularity of the requested location, and measurement technique.

The language for location-based predicates assumes that each user, who is unknown to the location server responsible for location measurements, is univocally identified via a user identifier (UID). For instance, a typical UID for location-based applications is the SIM number linking the user's identity to a mobile terminal. A unique identifier is also associated with each object, and any physical and/or moving entity that may need to be located (e.g., a vehicle with an on-board GPRS card). Moreover, to simplify the specification of location-based conditions, a set of map regions identified either via a geometric model (i.e., a range in a n-dimensional coordinate space) or a symbolic model (i.e., with reference to entities of the real world such as streets, cities, or buildings) are assumed to be predefined in the system [25]. For instance, let alice be a user identifier, and Manhattan_NY and University_Campus_Secretary be two map

Table 1. Examples of location-based predicates

Type	Predicate	Description
Position	inarea(*user, area*)	Evaluate whether *user* is located within *area*.
	disjoint(*user, area*)	Evaluate whether *user* is outside *area*.
	distance(*user, entity, min_dist, max_dist*)	Evaluate whether distance between *user* and *entity* is within interval [*min_dist, max_dist*].
Movement	velocity(*user, min_vel, max_vel*)	Evaluate whether *user*'s speed falls within range [*min_vel, max_vel*].
Interaction	density(*area, min_num, max_num*)	Evaluate whether the number of users currently in *area* falls within interval [*min_num, max_num*].
	local_density(*user, area, min_num, max_num*)	Evaluate the density within a 'relative' area surrounding *user*.

regions. Predicate inarea(alice,Manhattan_NY) requests alice to be located in Manhattan_NY; predicate velocity(alice,0,10) requests alice to be (semi-)static (speed included in [0,10]).

Besides location-based information, users and objects may be characterized by other properties that, for simplicity, are assumed to be stored in a profile, and to be referenced via the usual dot notation. For instance, alice.address indicates the address of user alice. Here, alice is the identity of the user (and therefore the identifier of the corresponding profile), and address is the name of the property. Also, since policies may need to refer to the user and object of the request being evaluated without need of introducing variables in the language, two keywords are used: **user**, which indicates the identifier of the requester, and **object**, which indicates the identifier of the object to which access is requested.

Location-based access control policies can then enrich the expressive power of current languages by allowing the evaluation of location-based conditions in the context of subject/object expressions. This way authorizations can result applicable to some access depending on conditions, such as, the location of the requester or of the resource.

Example 1. Consider a company responsible for the management of a mobile network that needs both strong authentication methods and expressive access control policies. Suppose that the Mobile Network Console (MNC) is the software that permits to reconfigure the mobile network. Table 2 presents some examples of protection requirements for such a service [26]. Managing a nation-wide mobile network is an extremely critical activity because reconfiguration privileges must be granted to strictly selected personnel only, that is, the execution of the MNC must be allowed according to high security standards. To this aim, Rule 1 states that only registered administrators that are static and alone in the server room can execute the MNC. In addition to the MNC execution privileges, also the access to mobile network's databases must be managed carefully and according to different security standards, depending on the level of risk of the data to be accessed. Access to logging and billing data is critical, because they include

Table 2. Examples of access control rules regulating access to the Mobile Network Console and databases of a mobile network

	subject		object	actions
	generic conditions	location-based conditions		
1	equal(user.role,admin) ∧ valid(user.username, user.password)	inarea(user.sim, Server_Room) ∧ density(Server_Room, 1, 1) ∧ velocity(user.sim, 0, 3)	equal(object.name,MNC)	execute
2	equal(user.role,admin) ∧ valid(user.username, user.password)	inarea(user.sim, Inf._System_Dept.) ∧ local_density(user.sim, Close_By, 1, 1) ∧ velocity(user.sim, 0, 3)	equal(object.category, Log&Bill)	read
3	equal(user.role,CEO) ∧ valid(user.username, user.password)	local_density(user.sim, Close_By, 1, 1) ∧ inarea(user.sim, Corp._Main_Office) ∧ velocity(user.sim, 0, 3)	equal(object.category, customer)	read
4	equal(user.role,CEO) ∧ valid(user.username, user.password)	local_density(user.sim, Close_By, 1, 1) ∧ disjoint(user.sim, Competitor_Location)	equal(object.category, StatData)	read
5	equal(user.role,guest) ∧ valid(user.username, user.password)	local_density(user.sim, Close_By, 1, 1) ∧ inarea(user.sim, Corporate_Location)	equal(object.category, StatData)	read

information about the position and movements of mobile operator's customers. Rule 2 is then defined and permits registered administrators that do not have other users in their proximity, static, and located in the information system department, to read logging and billing data. Access to customer-related information is usually less critical but still has to be handled in a highly secured environment and has to be granted only to selected personnel. Rule 3 states that registered CEOs that do not have other users in their proximity, static, and located in the corporate main office can read customer data. Finally, while statistical data about the network's operation is at a lower criticality level, access to them must be controlled, e.g., by preventing disclosure to competitors. To this aim, Rules 4 and 5 are defined: Rule 4 states that registered CEOs that do not have other users in their proximity and that are not located in a competitor location can read statistical data; Rule 5 states that registered guests that do not have other users in their proximity and located in the corporate location can read statistical data.

3.1.3   Location-Based Conditions Evaluation and Enforcement

The introduction of location-based conditions changes the usual way in which access control policies are evaluated and enforced. In particular, an ad-hoc solution must be designed to fully address both uncertainty and time-dependency of location-based information. The solution presented in [23] is based on two semantically uniform SLA parameters, *confidence* and *timeout*, returned by a LP to the ACE in response to the evaluation of a boolean query. Before illustrating how the access control process operates, we need to solve a basic problem: location-based predicates appear in rules as parts of a boolean formula (see Table 2), while the responses to boolean location queries are in the form of a triple [*bool_value,confidence,timeout*]. Then, to process a response from the Location Provider, the Access Control Engine will need to assign a truth value to the response. Intuitively, the transformation of a location predicate value into a boolean one requires the Access Control Engine to determine whether or not the value returned by the Location Provider can be considered valid for the purpose of controlling access. Such an evaluation will depend on parameters timeout and confidence returned by the Location Provider. Responses with a timeout that has already expired automatically trigger the re-evaluation of the predicate regardless of the other parameter values because considered as unreliable for any decision. Responses with a timeout that has yet not expired are evaluated with respect to the confidence value. The confidence value is compared with a *lower* and *upper* thresholds, specified for each location predicate. According to the result of this comparison (i.e., whether the confidence value is greater than the upper threshold, less than the lower threshold, or between the two), the boolean value contained in the response to a boolean query will be treated differently. More precisely, for each predicate in Table 1, an *Extended Truth Table* (ETT) defines a *lower* and *upper* thresholds, and a *MaxTries* parameter. If the *confidence* level for a given predicate evaluation is greater than the preset upper threshold, then *bool_value* returned by the LP is confirmed. If the *confidence* level is below the lower threshold, the location-based condition is evaluated to ¬*bool_value*. Otherwise, if the confidence level is between lower and upper thresholds neither the returned value nor its negation can be considered sufficiently reliable. Predicate re-evaluation is then triggered at the LP. In this case, the predicate is re-evaluated, at most *MaxTries* times, until the returned relevance is not between the thresholds. If after *MaxTries* re-evaluations of the predicate the outcome remains unchanged, the location-based condition evaluates *Undefined*.

Example 2. Suppose that for inarea predicate the lower and upper thresholds are 0.2 and 0.8, respectively, and that
inarea(Alice,Manhattan_NY) = [*True*,*0.85*,*2009-01-20 9:00pm*]
is the triple returned by the LP to the ACE stating that Alice is located in Manhattan_NY with *confidence* of 85%. Such an assessment is to be considered valid until *9:00pm* of *January 20th, 2009*. The ACE evaluates inarea(Alice,Manhattan_NY) to *True*, since *0.85>0.80*.

3.2   GeoXACML

The *Geospatial eXtensible Access Control Markup Language* (GeoXACML) [24] has been introduced by the Open Geospatial Consortium (OGC) as an extension to the XACML Policy Language [11], to support the declaration and enforcement of predicates based on geographic information. GeoXACML, which becomes an OGC standard in February 2008, defines ad-hoc extensions to XACML for including geometric attributes and spatial functions (predicates). The attributes introduced are derived from the Geographic Markup Language (GML) and defined in the GeoXACML Core Geometry Model. Examples of geometric attributes are: *Point*, that models a single location; *LineString*, that represents a curve with linear interpolation between Points; *Polygon*, that identifies a planar area defined by an exterior boundary, and zero or more interior boundaries; *MultiPoint, MultiLineString*, and *MultiPolygon*, that represent a collection of Points, LineStrings, and Polygons, respectively. The GeoXACML predicates can be partitioned into different categories: *topological, geometric, bag, conversion,*

Table 3. Examples of GeoXACML spatial functions

Type	Function	Description
Topological	Contains(*g1:Geometry*, *g2:Geometry*) : Boolean	Returns a true value if and only if geometry g2 lies in the closure (boundary union interior) of geometry g1.
	Crosses(*g1:Geometry*, *g2:Geometry*) : Boolean	Returns a true value if and only if geometries g1 and g2 have some but not all interior points in common, and the dimension of the intersection is less than that of both of the geometries.
	Disjoint(*g1:Geometry*, *g2:Geometry*) : Boolean	Returns a true value if and only if the geometries g1 and g2 have no points in common.
	Equals(*g1:Geometry*, *g2:Geometry*) : Boolean	Returns a true value if and only if geometries g1 and g2 are equal (geometrically contain exactly the same points).
	Overlaps(*g1:Geometry*, *g2:Geometry*) : Boolean	Returns a true value if and only if geometries g1 and g2 have some but not all points in common, and the intersection has the same dimension as each geometry.
	Within(*g1:Geometry*, *g2:Geometry*) : Boolean	Returns a true value if and only if geometry g1 is spatially within geometry g2, that is, if every point on g1 is also on g2.
Geometric	Boundary(*g:Geometry*) : Bag	Returns a bag of geometry values representing the combinatorial boundary of geometry g.
	Centroid(*g:Geometry*) : Geometry	Returns the point that is the geometric center of gravity of the geometry g.
	Intersection(*g1:Geometry*, *g2:Geometry*) : Bag	Returns a bag of geometry values representing the Point set intersection of geometry g1 and geometry g2.
	Union(*g1:Geometry*, *g2:Geometry*) : Bag	Returns a bag of geometry values representing the Point set union of geometry g1 with geometry g2.
	Area(*g:Geometry*) : Double	Returns a value representing the area of geometry g.
	Distance(*g1:Geometry*, *g2:Geometry*) : Double	Returns a value representing the shortest distance in meter between any two points in the two geometries g1 and g2.

and *set*. Table 3 presents some predicates, which can be used for testing topological relations between geometries (we refer to the OGC proposal [24] for the complete set of predicates). A geometry provides a description of geographic characteristics (e.g., locations, shapes). The encoding of geometry depends on the coordinate reference system (CRS) or spatial reference system (SRS) that is used. It is important to note that some predicates provides supporting functionalities only. For instance, the predicates in the conversion category assist in the conversion of other measurement units in terms of meters or square meters (the only accepted by GeoXACML). The use of these conversion functions should however be minimized to avoid unnecessary delays in information processing. Another set of predicates providing supporting functionality, included in the geometric category, is used to verify special characteristics of geometries. For instance, to verify whether a geometry has anomalous geometric points (e.g., self intersection, or self tangency).

GeoXACML, being an extension of XACML, has the same policy syntax of XACML. A GeoXACML policy is then composed of a set of `Rule` elements, each one leading to a binary effect (i.e., *Permit* or *Deny*). An authorization

```
<Rule ... Effect=''Permit'' RuleId=''Example''>
  <Target>
    <Subjects>
      <Subject>
        <SubjectMatch MatchId=''urn:oasis:names:tc:xacml:1.0:function:string-equal''>
          <AttributeValue DataType=''http://www.w3.org/2001/XMLSchema#string''>
            John Brown
          </AttributeValue>
          <SubjectAttributeDesignator
          DataType=''http://www.w3.org/2001/XMLSchema#string''
          AttributeId=''urn:oasis:names:tc:xacml:1.0:subject:subject-id''/>
        </SubjectMatch>
      </Subject>
    </Subjects>
    <Resources>
      <Resource>
        <ResourceMatch MatchId=''urn:oasis:names:tc:xacml:1.0:function:string-equal''>
          <AttributeValue DataType=''http://www.w3.org/2001/XMLSchema#string''>
            Building
          </AttributeValue>
          <AttributeSelector
          RequestContextPath=''name(//ca:CityModel/gml:featureMember/ca:Building[1]''
          DataType=''http://www.w3.org/2001/XMLSchema#string''/>
        </ResourceMatch>
      </Resource>
    </Resources>
    <Actions>
      <Action>
        <ActionMatch MatchId=''urn:oasis:names:tc:xacml:1.0:function:string-equal''>
          <AttributeValue DataType=''http://www.w3.org/2001/XMLSchema#string''>
            Read
          </AttributeValue>
          <ActionAttributeDesignator
          AttributeId=''urn:oasis:names:tc:xacml:1.0:action:action-id''
          DataType=''http://www.w3.org/2001/XMLSchema#string''/>
        </ActionMatch>
      </Action>
    </Actions>
  </Target>
  <Condition>
    <Apply FunctionId=''urn:oasis:names:tc:xacml:1.0:function:all-of''>
      <Function
      FunctionId=''urn:oasis:names:tc:xacml:1.0:function:string-equal''/>
        <AttributeValue
        DataType=''http://www.w3.org/2001/XMLSchema#string''>Wincott Street</AttributeValue>
        <AttributeSelector ''
        RequestContextPath=''//ca:CityModel/gml:featureMember/ca:Building/ca:address''
        DataType=''http://www.w3.org/2001/XMLSchema#string''>
    </Apply>
  </Condition>
</Rule>
```

Fig. 2. An example of GeoXACML rule

decision is derived by first determining all the rules applicable to a given request. All matching rules are then combined according to a predefined algorithm to obtain the resulting effect of the policy. When more policies are applicable, all resulting policy effects produced for a given request must be combined to produce the final authorization decision. The main difference between XACML and GeoXACML is that the latter supports the declaration of spatial restrictions, which are expressed through the predicates above-mentioned. Figure 2 shows an example of GeoXACML rule whose Effect is Permit. For simplicity, namespaces in the rule element are omitted. The rule's target (i.e., element Target) has three main elements: Subjects, which defines the rule's subjects, that is, *John Brown*; Resources, which identifies the rule's objects, that is, *Building*; and Actions, which specifies the actions that can be performed, that is, *Read*. Element Condition introduces further matching conditions; in our example, the fact that address has to be *Wincott Street*. The semantic of the rule is that "the user *John Brown* can *Read* the information object of class *Building*, if the address is *Wincott Street*" [27].

4   Protecting Location Information in Ubiquitous Computing

Today ubiquitous technologies give the basis for accessing, mining, and managing large amount of location information. Such information, however, can be extremely sensitive, and lack of its protection can result in several attacks to the user's personal sphere. Research has been approaching the problem of protecting access to location information from different perspectives, such as the development of enhanced access control architectures or the definition of new languages for protecting location information. In the following, we illustrate some of these proposals.

4.1   Geopriv

IETF *Geopriv* working group [28] proposes a solution for protecting privacy of location information, when it is transmitted and shared over the Internet. Geopriv's main principles and considered threats have been formalized in the IETF RFC 3693 and RFC 3694 [29,30]. Geopriv considers a scenario in which a requester asks for location information of a target to a location server. An architecture to manage such a scenario has been introduced and includes four main parties.

- *Location Generator* (LG) gathers location information of users and makes it available to the Location Server.
- *Location Server* (LS) provides location services to Location Recipients, and stores the location information of the users.
- *Location Recipient* (LR) subscribes for a location-based service provided by the LS, and requests access to the location information stored by the LS.

– *Rule Holder/Maker* (RH/M) defines the privacy policies which regulate the disclosure of location information to the LR. The policies are enforced by the LS.

Based on these logical components, different architectural layouts are possible. For instance, LG and LS may coexist on the same mobile device (e.g., a GPS receiver) or could be distributed components communicating remotely. The RH/M could be a centralized component managing privacy rules and communicating them to the corresponding LS, or it could be co-located with the LS.

The location information of users is part of a container, called *Location Object* [31]. In addition to the location information, a location object includes privacy preferences (i.e., usage-rules), that remain attached to the location information for its entire life-cycle. In particular, usage-rules allow the definition of conditions that can: *i)* limit retransmission (e.g., "retransmission-allowed"), *ii)* limit retention (e.g., "retention-expires" date), and *iii)* contain a reference to external rulesets.

Geopriv IETF RFC 4745 [32] defines the framework for creating privacy policies that regulate the release of location information. A Geopriv privacy policy, encoded in XML, is composed of a `ruleset` element that contains an unordered list of `rules` elements corresponding to positive authorizations. Each rule has an element `conditions`, `actions`, and `transformations`. The `condition` element is a set of expressions, each of which evaluates to either true or false. A limited set of conditions can be specified in the `conditions` element: `identity`, `sphere`, and `validity`. The `identity` element restricts the rule matching either to a single identity, using the `one` element, or a group of identities, using the `many` element. In particular, the `one` element identifies exactly one authenticated entity or user, while the `many` element represents a generic number of users in a domain (i.e., it matches the domain part of an authenticated identity). Moreover, the `identity` element can exclude individual users or users belonging to a specific domain through the `except` element. The `sphere` element can be used to match the state (e.g., work, home) a target holds at the time of the access request evaluation. Finally, the `validity` element is used to restrict the time validity of each rule. Additional condition elements can be added by proposing extensions to the privacy policy specification in RFC 4745. The `actions` element specifies actions to be applied before the release of location information. The `transformations` element specifies how the location information should be modified when a permission is granted; for instance, it can state that the original location should be made less precise. While conditions can be considered as the 'if'-part of the rules, which states whether the rule is applicable, actions and transformations form the 'then'-part, which determines the operations to be performed before disclosing information.

Figure 3 shows an example of Geopriv rule. The rule states that, during February 2009, the authenticated entity *sip:bob@example.com* or *mailto:dave@example.net* can access the location information, protected by the rule, if target's sphere is equal to "work".

```
<rule id=''a7k55r7''>
  <conditions>
    <identity>
      <one id=''sip:bob@example.com''/>
      <one id=''mailto:dave@example.net''/>
    </identity>
    <sphere value=''work''/>
    <validity>
      <from>2009-02-01T00:00:00.000-03:00</from>
      <until>2009-02-28T23:59:59.000-03:00</until>
    </validity>
  </conditions>
  <actions/>
  <transformations/>
</rule>
```

Fig. 3. An example of Geopriv rule

4.2   Protecting Location Information in Mobile Applications

Different works have addressed the problem of protecting location information in mobile applications.

A first line of research focuses on extending Platform for Privacy Preferences (P3P) for protecting the secondary uses of location information [33,34,35]. P3P [36,37] has been originally designed by the World Wide Web Consortium (W3C) to address the need of users to assess whether the privacy practices defined by a Web site comply with their privacy requirements, before the release of personal information. Privacy requirements are expressed through *A P3P Preference Exchange Language* (APPEL) [38]. Starting from the work done in P3P, Langheinrich [34] proposes a privacy awareness system (pawS) for ubiquitous and pervasive environments, where services collect users data. The main goal of pawS is to provide an infrastructure that allows users to protect their privacy and to keep track of all data released and of their subsequent management at the service side. pawS uses P3P to encode data usage policies of the service and users define their preferences through APPEL. In pawS, a mobile user carries a mobile device with a *privacy assistant*. When a user enters a geographical area in which a number of services are available (e.g., location tracking service using video-camera), the privacy assistant is prompted with the data collection practices of the service. This communication happens on wireless channels. To save the battery of the portable devices and make the system appealing also for mobile users, data usage practices are delegated by the user to a *personal privacy proxy* residing on the Internet, which is responsible for managing all negotiations with the service. In particular, the personal privacy proxy asks the *service proxy* for service policies and then matches them with the users' preferences. If the matching is successful, the service is used and data released, otherwise the service is disabled. Access control policies exploiting the location of the requesters are evaluated and enforced. Myles et al. [35] discuss a location-based scenario where applications require location information of the users for service release. The main goal is, on the one side, to balance the need of mechanisms to protect users' privacy limiting service intrusiveness, and, on the other side, to minimize the overhead given to the users. The proposed system architecture is composed of three main entities: *i)* a location server, that manages positioning systems (e.g., GPS, cellular technologies) and answers to requests for location information; *ii)* several validators, that are responsible for evaluating the requests and determining whether the location information can be released, based on preferences of the users; *iii)* client applications, that submit requests for location information. The authors assume trust relationships between users, validators, and location servers. Users are registered with at least one location server and store their requirements within it. These requirements are implemented by the validators. When a client application needs to access the location of a user, it first selects the relevant location server, and then submits the request. Such a request also includes the privacy policies that specify how the client application will manage the data after their collection. The privacy policies can be expressed through an extension of P3P that allows the modeling of requests initiated by the application. After receiving the request, the privacy policies are matched with the privacy preferences stored by the validators. Such preferences can contain restrictions based on the time of the request and on geographical areas. Validators can implement a variety of mechanisms for privacy preferences specifications (e.g., APPEL). Hong et al. [33] provide an extension to P3P for representing user privacy preferences in context-aware applications. The authors add features to the P3P language to express the identifiers of the users whose locations are collected, the time period in which the data can be accessed, and the location from which the data can be managed. They propose a centralized architecture that includes a middleware responsible for matching preferences and policies. The middleware is enriched with a plug-in service to support context-aware applications, called *privacy database mediator*. The privacy database mediator provides functionality to automatically generate privacy policies and user preferences according to the context.

Another line of research has addressed the definition of authorization architectures, based on certificates and encryption, to protect location information. Hauser and Kabatnik [39] address the problem of protecting the location information of the users by providing a privacy-aware architecture that allows users to define rules regulating the access to their location information. The proposed solution relies on asymmetric encryption and authorization certificates. The requester asks the location server for the position of a given target (position query), by sending the authorization certificate released by the target. The certificate is a chipertext encrypted with the public key of the location server and contains the pseudonym of the target. The location server, after decrypting the chipertext, retrieves the target's pseudonym, and satisfies the subject request by releasing the target's position. Note that the location server is not aware of the real identity of the targets. A more complex solution is also provided for queries that ask for a list of targets in a given area. In this case, a certificate specifying the privilege to query a specific area is not enough, but rather the requester has to send the authorizations of all the users relevant for the query. Hengartner and Steenkiste [40,41] use digital certificates combined with rule-based policies to protect location information. They consider an environment in which users submit requests to a "people locator", which in turn collects the relevant location information through multiple positioning systems. The authors propose an access control mechanism where policies are encoded as digital certificates using SPKI/SDSI. Location policies can specify the entities that can access the location information, the granularity of the information returned to the requester, the location of the requester, and the time allowed for each access. In case of forwarded requests, *trust policies* are used to verify whether the intermediate service is trusted or not to forward a request and receive a response. Finally, *delegation* of right is allowed to grant access to other entities. Atluri and Chun [42] present Geo-Spatial Data Authorization Model (GSAM), an authorization model that protects access to geospatial data. GSAM provides policies evaluating geospatial and temporal characteristics of user's credentials and data objects, and introduces different types of actions (e.g., zoom-in, view, and download). For instance, GSAM defines security and privacy policies that allow access to low resolution images regardless of location coordinates of users, whereas restrict access to high resolution images only for those users located in a particular region.

5   Open Issues

We briefly describe some open issues that need to be taken into consideration in the future development of access control systems for location-based services.

- *Reliable enforcement based on fine-grained context information.* As discussed, a key aspect to the success of location-based access control systems is the definition of a reliable enforcement solution, able to verify information which is approximate and time-variant. In the near future, location servers will provide a wealth of additional environment-related knowledge (e.g., is the user sitting at her desk or walking toward the door? Is she alone or together with others?), that may give the opportunity of defining and evaluating new classes of location-based conditions in the context of LBAC systems. LBAC systems however may be flawed by the intrinsic errors of location measurements, in calculating such fine-grained knowledge. Future access control mechanisms should then try to enhance current approaches to the management of uncertain information, thus providing policy evaluation mechanisms able to support fine-grained location information.
- *Privacy-aware LBAC.* An important aspect to consider in today access control systems is the protection of the user privacy. Some solutions have been presented in the past (e.g., [9]) which provide, on the one side, access control functionality and, on the other side, privacy protection. However, LBAC systems introduce new threats that should be carefully considered. In particular, a fundamental issue to be addressed considers the conflicting requirements of preserving users privacy and of providing high quality LBAC. A suitable protocol should in fact balance the tradeoff between the level of location accuracy requested by LBAC providers and the protection of the location

information requested by the users. A possible approach in developing a privacy-aware LBAC may integrate access control with location privacy solutions (e.g., obfuscation [6,7] and anonymity [43,44,45,46] techniques).

– *Integration of different location sources.* An important issue in the development of LBAC systems is represented by the availability of several location servers, which support different positioning systems for measuring location of the users. In this context, a solution which implements communication and negotiation protocols between the LBAC system and multiple, functionally equivalent, location servers is needed. These protocols should provide an approach based on service level agreement attributes which maximize the QoS and/or cost/benefit functions.

6   Conclusions

In this chapter, we discussed how the advent of location-based services and the availability of precise location information are changing traditional access control systems. We considered two different scenarios: *i)* the definition of a location-based access control system, which integrates, evaluates, and enforces traditional access control policies enriched with conditions based on the physical position of users; *ii)* the development of enhanced access control systems for protecting the location information. For both of them, we investigated recent proposals and ongoing work. Finally, we presented open issues that need further investigation.

Acknowledgments

This work was supported in part by the EU, within the 7th Framework Programme (FP7/2007-2013) under grant agreement no. 216483 "PrimeLife".

References

1. Varshney, U.: Location management for mobile commerce applications in wireless internet environment. ACM Transactions on Internet Technology (TOIT) 3(3), 236–255 (2003)
2. Ardagna, C., Cremonini, M., Damiani, E., De Capitani di Vimercati, S., Samarati, P.: Privacy-enhanced location services information. In: Acquisti, De Capitani di Vimercati, Gritzalis, Lambrinoudakis (eds.) Digital Privacy: Theory, Technologies and Practices. Auerbach Publications (2007)
3. Enhanced 911: Wireless Services, http://www.fcc.gov/911/enhanced/
4. Chicago Tribune: Rental firm uses GPS in speeding fine, p.9. Associated Press, Chicago (July 2, 2001)
5. Duckham, M., Kulik, L.: Location privacy and location-aware computing. In: Drummond, J., Billen, R., Forrest, D., Joao, D. (eds.) Dynamic & Mobile GIS: Investigating Change in Space and Time, pp. 34–51. CRC Press, Boca Raton (2006)
6. Ardagna, C., Cremonini, M., Damiani, E., De Capitani di Vimercati, S., Samarati, P.: Location privacy protection through obfuscation-based techniques. In: Barker, S., Ahn, G.-J. (eds.) Data and Applications Security 2007. LNCS, vol. 4602, pp. 47–60. Springer, Heidelberg (2007)

7. Duckham, M., Kulik, L.: A formal model of obfuscation and negotiation for location privacy. In: Gellersen, H.-W., Want, R., Schmidt, A. (eds.) Pervasive 2005. LNCS, vol. 3468, pp. 152–170. Springer, Heidelberg (2005)

8. Samarati, P., De Capitani di Vimercati, S.: Access control: Policies, models, and mechanisms. In: Focardi, R., Gorrieri, R. (eds.) FOSAD 2000. LNCS, vol. 2171, p. 137. Springer, Heidelberg (2001)

9. Ardagna, C., Cremonini, M., De Capitani di Vimercati, S., Samarati, P.: A privacy-aware access control system. Journal of Computer Security 16(4), 369–392 (2008)

10. De Capitani di Vimercati, S., Foresti, S., Samarati, P.: Recent advances in access control. In: Gertz, M., Jajodia, S. (eds.) Handbook of Database Security: Applications and Trends, Springer, Heidelberg (2008)

11. eXtensible Access Control Markup Language (XACML): Version 2.0 (February 2005), http://docs.oasis-open.org/xacml/2.0/access_control-xacml-2.0-core-spec-os.pdf

12. Bonatti, P., Samarati, P.: A unified framework for regulating access and information release on the web. Journal of Computer Security 10(3), 241–272 (2002)

13. Akyildiz, I., Ho, J.: Dynamic mobile user location update for wireless PCS networks. Wireless Networks, vol. 1 (1995)

14. Faria, D., Cheriton, D.: No long-term secrets: Location-based security in overprovisioned wireless lans. In: Proc. of the 3rd ACM Workshop on Hot Topics in Networks (HotNets-III), San Diego, CA, USA (November 2004)

15. Garg, S., Kappes, M., Mani, M.: Wireless access server for quality of service and location based access control in 802.11 networks. In: Proc. of the 7th IEEE Symposium on Computers and Communications (ISCC 2002), Taormina/Giardini Naxos, Italy (July 2002)

16. Myllymaki, J., Edlund, S.: Location aggregation from multiple sources. In: Proc. of the 3rd IEEE International Conference on Mobile Data Management (MDM 2002), Singapore (January 2002)

17. Cho, Y., Bao, L., Goodrich, M.: Secure access control for location-based applications in WLAN systems. In: Proc. of the 3rd IEEE International Conference on Mobile Adhoc and Sensor Systems, Vancouver, Canada (October 2006)

18. Nord, J., Synnes, K., Parnes, P.: An architecture for location aware applications. In: Proc. of the 35th Hawaii International Conference on System Sciences, Hawaii, USA (2002)

19. Sastry, N., Shankar, U., Wagner, S.: Secure verification of location claims. In: Proc. of the ACM Workshop on Wireless Security (WiSe 2003), San Diego, CA, USA (September 2003)

20. Zhang, G., Parashar, M.: Dynamic context-aware access control for grid applications. In: Proc. of the 4th International Workshop on Grid Computing (Grid 2003), Phoenix, AZ, USA (November 2003)

21. Atallah, M., Blanton, M., Frikken, K.: Efficient techniques for realizing geo-spatial access control. In: Proc. of the 2nd ACM Symposium on InformAtion, Computer and Communications Security (ASIACCS 2007), Singapore (March 2007)

22. Atluri, V., Shin, H., Vaidya, J.: Efficient security policy enforcement for the mobile environment. Journal of Computer Security 16(4), 439–475 (2008)

23. Ardagna, C., Cremonini, M., Damiani, E., De Capitani di Vimercati, S., Samarati, P.: Supporting location-based conditions in access control policies. In: Proc. of the ACM Symposium on InformAtion, Computer and Communications Security (ASIACCS 2006), Taipei, Taiwan (March 2006)

24. Open Geospatial Consortium: Geospatial eXtensible Access Control Markup Language (GeoXACML) Version 1.0 (February 2008),
    http://portal.opengeospatial.org/
25. Marsit, N., Hameurlain, A., Mammeri, Z., Morvan, F.: Query processing in mobile environments: a survey and open problems. In: Proc. of the 1st International Conference on Distributed Framework for Multimedia Applications (DFMA 2005), Besancon, France (February 2005)
26. Ardagna, C., Cremonini, M., De Capitani di Vimercati, S., Samarati, P.: Privacy-enhanced location-based access control. In: Gertz, M., Jajodia, S. (eds.) The Handbook of Database Security: Applications and Trends, Springer, Heidelberg (2007)
27. Matheus, A.: Declaration and Enforcement of Access Restrictions for Distributed Geospatial Information Objects. Ph.D Thesis (2005)
28. Geographic Location/Privacy (geopriv),
    http://www.ietf.org/html.charters/geopriv-charter.html
29. Cuellar, J., Morris, J., Mulligan, D., Peterson, J., Polk, J.: Geopriv Requirements. IETF RFC 3693 (February 2004)
30. Danley, M., Mulligan, D., Morris, J., Peterson, J.: Threat Analysis of the Geopriv Protocol. IETF RFC 3694 (February 2004)
31. Cuellar, J.: A Presence-based GEOPRIV Location Object Format. IETF RFC 4119 (December 2005)
32. Schulzrinne, H., Tschofenig, H., Morris, J., Cuellar, J., Polk, J., Rosenberg, J.: Common Policy: A Document Format for Expressing Privacy Preferences. IETF RFC 4745 (February 2007)
33. Hong, D., Yuan, M., Shen, V.Y.: Dynamic privacy management: a plug-in service for the middleware in pervasive computing. In: Proc. of the 7th International Conference on Human Computer Interaction with Mobile Devices & Services (MobileHCI 2005), Salzburg, Austria (2005)
34. Langheinrich, M.: Privacy by design-principles of privacy-aware ubiquitous systems. In: Abowd, G.D., Brumitt, B., Shafer, S. (eds.) UbiComp 2001. LNCS, vol. 2201, p. 273. Springer, Heidelberg (2001)
35. Myles, G., Friday, A., Davies, N.: Preserving privacy in environments with location-based applications. IEEE Pervasive Computing 2(1), 56–64 (2003)
36. Cranor, L.: Web Privacy with P3P. O'Reilly & Associates, Sebastopol (2002)
37. World Wide Web Consortium (W3C): Platform for privacy preferences (P3P) project (April 2002), http://www.w3.org/TR/P3P/
38. World Wide Web Consortium (W3C): A P3P Preference Exchange Language 1.0 (APPEL1.0) (April 2002), http://www.w3.org/TR/P3P-preferences/
39. Hauser, C., Kabatnik, M.: Towards Privacy Support in a Global Location Service. In: Proc. of the IFIP Workshop on IP and ATM Traffic Management (WATM/EUNICE 2001), Paris, France (September 2001)
40. Hengartner, U., Steenkiste, P.: Protecting access to people location information. Security in Pervasive Computing (March 2003)
41. Hengartner, U., Steenkiste, P.: Implementing access control to people location information. In: Proc. of the ACM Symposium on Access Control Models and Technologies 2004 (SACMAT 2004), Yorktown Heights, New York, USA (2004)
42. Atluri, V., Chun, S.: An authorization model for geospatial data. IEEE Transactions on Dependable and Secure Computing 1(4), 238–254 (2004)
43. Bettini, C., Wang, X., Jajodia, S.: Protecting privacy against location-based personal identification. In: Jonker, W., Petković, M. (eds.) SDM 2005. LNCS, vol. 3674, pp. 185–199. Springer, Heidelberg (2005)

44. Ghinita, G., Kalnis, P., Skiadopoulos, S.: Privè: Anonymous location-based queries in distributed mobile systems. In: Proc. of the International World Wide Web Conference (WWW 2007), Banff, Canada (May 2007)
45. Gruteser, M., Grunwald, D.: Anonymous usage of location-based services through spatial and temporal cloaking. In: Proc. of the 1st International Conference on Mobile Systems, Applications, and Services, San Francisco, CA, USA (May 2003)
46. Mokbel, M., Chow, C.Y., Aref, W.: The new casper: Query processing for location services without compromising privacy. In: Proc. of the 32nd International Conference on Very Large Data Bases, Seoul, Korea (September 2006)

Location Privacy in RFID Applications

Ahmad-Reza Sadeghi[1], Ivan Visconti[2], and Christian Wachsmann[1]

[1] Horst Görtz Institute for IT-Security (HGI)
Ruhr-University Bochum, Germany
{ahmad.sadeghi,christian.wachsmann}@trust.rub.de
[2] Dipartimento di Informatica ed Applicazioni
University of Salerno, Italy
visconti@dia.unisa.it

Abstract. RFID-enabled systems allow fully automatic wireless identification of objects and are rapidly becoming a pervasive technology
with various applications. However, despite their benefits, RFID-based
systems also pose challenging risks, in particular concerning user privacy. Indeed, improvident use of RFID can disclose sensitive information
about users and their locations allowing detailed user profiles. Hence,
it is crucial to identify and to enforce appropriate security and privacy
requirements of RFID applications (that are also compliant to legislation). This chapter first discusses security and privacy requirements for
RFID-enabled systems, focusing in particular on location privacy issues.
Then it explores the advances in RFID applications, stressing the security
and privacy shortcomings of existing proposals. Finally, it presents new
promising directions for privacy-preserving RFID systems, where as a case
study we focus electronic tickets (e-tickets) for public transportation.

1 Introduction

Radio frequency identification (RFID) is a technology that enables RFID *readers*
to perform fully automatic wireless identification of objects that are labeled with
RFID *tags*. Initially, this technology was mainly used for electronic labeling of
pallets, cartons, or products to enable seamless supervision of supply chains.
Today, RFID technology is widely deployed to many other applications as well,
including animal identification [1], library management [2], access control [1,
3, 4, 5], electronic tickets [3, 4, 5], electronic passports [6], and even human
implantation [7].

Security and privacy risks of RFID. The most deterrent risk of RFID systems are
tracing attacks that aim at obtaining user-related data, including user identities
and user locations. THis information provides profiles on users' personal habits,
interests, preferences, and even on their health and financial status. For instance,
frequent visits to hospitals may indicate health problems. Due to the wireless
interface of RFID, user-related information may be leaked unnoticeably by the
RF interface to unauthorized entities (i.e., those that are not trusted by the user).
Thus, an important security objective of an RFID system is to ensure location

C. Bettini et al. (Eds.): Privacy in Location-Based Applications, LNCS 5599, pp. 127–150, 2009.

privacy by preventing unauthorized access to user-related data (*confidentiality*), unauthorized identification of users (*anonymity*), as well as unauthorized tracing of tags by linking their communication (*unlinkability*).

Despite these privacy risks, classical threats to authentication and identification systems must be considered as well. Indeed, potential threats to RFID systems are attacks, where an adversary tries to impersonate or clone a legitimate tag. By legitimate we mean a tag created by an accredited tag issuer. Thus appropriate countermeasures must be provided (*authentication* and *unclonability*). However, there are some other risks such as *denial-of-service* attacks, where an adversary unnoticeably interacts with tags and exploits deficiencies of the underlying protocols to permanently disable legitimate tags [8], which must also be prevented (*availability*).

Security and privacy models for RFID. It is essential to carefully formalize the security and privacy goals discussed above to enable the design of provably secure and privacy-preserving RFID protocols that are also usable in practice. Existing literature proposes several security and privacy models for RFID. One of the most mature models has been presented in [9]. This model provides a game-based definition for privacy and security for RFID that generalizes and improves many previous models, including [8, 10, 11, 12, 13]. As pointed out in Section 2.5, these models have limitations in the modeling of location privacy since they either do not consider adversaries with access to side-channel information (e.g., information on whether authentication of a tag was successful or not) or do not consider privacy of tags that have been compromised by (i.e., whose secrets have been revealed to) an adversary.

Privacy-preserving protocols for RFID. Currently available RFID tags at most support random number generation and symmetric-key cryptography [1, 4] while the support for public-key cryptography is still very expensive.

A general problem concerning privacy-preserving authentication of tags that are limited to symmetric-key cryptography is how to inform the reader which key should be used for authentication. Indeed, a tag cannot disclose its identity before the reader has been authenticated since this would violate unlinkability. However, a reader cannot authenticate a tag unless it knows the identity (i.e., the key) of that tag. There is a large body of literature proposing solutions to this problem (e.g., [2, 14, 15, 16, 17, 18, 19, 20, 21, 22, 23, 24, 25]), however, almost all of them have deficiencies for their deployment in real-world applications. For instance, in many protocols the computational effort of a reader to verify a tag depends on the total number of legitimate tags in the system, which is unacceptable for systems with a huge amount of tags (e.g., electronic product labels or e-tickets). Other protocols require the reader to have a permanent online connection to some trusted database, which is inappropriate for systems that require mobility of readers (e.g., mobile ticket inspectors in transit systems).

Applications. We will focus on electronic tickets (e-tickets) based on RFID. This application of RFID is becoming very popular among operators of public transportation and many proprietary RFID-based e-ticket systems are already widely deployed in practice. E-tickets offer several advantages including fast and convenient verification of tickets and aggravation of forgeries by cryptographic means. However, e-tickets also introduce several risks, in particular concerning the location privacy of users. The motivation for focusing on e-tickets is that known systems used in practice do not consider location privacy whereas the literature on privacy-preserving e-tickets and authentication of RFID devices (that could be applied to e-tickets) typically lacks practicality. Thus, the design and secure implementation of a privacy-preserving and usable e-ticket system based on RFID is currently an interesting open problem.

Organization. This chapter first discusses security and privacy requirements for RFID-enabled systems, focusing in particular on location privacy issues. Then it explores the advances in RFID applications, stressing the security and privacy shortcomings of existing proposals. Finally, it illustrates new promising directions for privacy-preserving RFID-enabled systems, considering electronic tickets (e-tickets) for public transportation as a case study.

2 RFID Systems and Their Requirements

In this section we describe the actors of an RFID system and we specify the requirements that such a system should satisfy in order to be considered secure.

2.1 RFID System Structure

A typical RFID system consists of many *tags* and at least one *reader* that is used to communicate with tags. A tag is an integrated circuit connected to an antenna, which both are usually integrated into some plastic card or sticker that is attached to the object to be identified. Currently available passive[1] RFID devices are powered by readers and thus cannot initialize communication, have limited memory, are not tamper-resistant and are limited to basic cryptographic computations, including keyed hashing, symmetric-key encryption and random number generation [1, 26]. Often, the purpose of the readers is to distinguish legitimate tags from unknown tags. In practice, RFID readers typically have secure (i.e., confidential and authenticated) access to some backend database that contains information on all legitimate tags (see Figure 1).

Today, RFID is mainly used for identification or authentication purposes including access control [7] or anti-counterfeiting systems [27]. Users of an RFID system own one or more tags that can be interrogated without optical or physical contact. This greatly enhances convenience in access control systems since users do not need to insert their security token into a reader but can leave it in their wallets or pockets. However, wireless interaction is imperceptible and thus may allow unauthorized entities to obtain user-related data including personal information and locations. As a consequence, in addition to the threats to conventional authentication systems, RFID must consider privacy and security problems that are related to the RF interface.

---

[1] Active RFID devices have an on-tag power supply and thus are too expensive and too big for most commercial applications.

Fig. 1. Typical RFID system architecture

2.2   Attacks on RFID

The main goal of every authentication scheme, including RFID, is to prevent unauthorized users from cheating any honest reader in order to obtain an unauthorized access. Beyond guaranteeing this main goal, there are some other, subtle attacks against RFID systems that do not only aim at making location profiles of users but are crucial for the deployment of RFID systems to real-world applications.

Impersonation. The most obvious attack against RFID systems is motivated by unauthorized entities. The adversary must obtain or simulate a tag that is accepted by an honest reader. To achieve this, the adversary may perform various attacks including man-in-the-middle or replay attacks against the underlying authentication protocols or he may attempt to create forged tags or to copy tags of honest users.

Tracing. A more subtle attack aims at obtaining information on users and their movements. When using conventional authentication protocols, a tag can be easily identified during verification, which enables readers to trace tags. Moreover, if users can be identified when obtaining a tag (e.g., when using an identifying payment method like credit cards for buying an RFID-enabled e-ticket), the issuer of the tag can link the corresponding tag to the identity of its owner. Since the issuer and the readers are typically under the control of the same entity (e.g., the transit enterprise in case of e-tickets), this results in a complete loss of the user's location privacy. For instance, the transit enterprise may link the transactions of the user's ticket and correlate this data with the geographical location of the readers. However, in this case, user information is managed by one single known entity that can be committed by law to the confidential use of the collected user data, and that can be monitored by means of inspections (similar observations hold for credit card companies).

Summing up, a primary goal of an RFID system is to ensure location privacy by preventing the disclosure of information on users and their movements to all entities that are not trusted by the users.

Denial-of-service. Another type of adversary may want to harm (e.g., to blackmail) the company running the RFID system by disturbing the authentication process of honest users. Besides financial losses such an attack would seriously damage the reputation of the affected company and thus should be prevented by every dependable RFID system. However, since tags are wireless devices that can be attacked unnoticeably, an adversary may try to exploit deficiencies of the protocols such that a tag is no longer accepted by honest readers.

Depending on the underlying use case and business model, RFID protocols must be carefully designed to prevent some or all of these attacks. Section 2.3, introduces different trust and adversary models for RFID systems. A complete list of requirements for practical privacy-preserving RFID systems is given in Section 2.4.

2.3   Trust and Adversary Models

In an ideal setting, no entity must be trusted. However, in practice, at least the issuer must be trusted to only create tags for eligible users. Moreover, each reader must be trusted to only accept tags that have been issued by a genuine issuer. These are reasonable assumptions since in practice, the issuing entity and the readers are typically physically controlled by the same entity (e.g., the transit enterprise in case of e-tickets) or share the same goals.

Ideally, users should be anonymous to every entity, including the tag issuer and all readers. However, due to technical restrains this is not always feasible in practice. Thus, a reasonable trust model for a practical solution is that users must at least trust the issuer and, depending on the implementation, also all readers. Obviously, a trust model which only requires the tag issuer to be trusted is preferable.

Summing up, while tag issuers and readers must trust each other, for users there are three possible trust models:

- TM1: Users do not need to trust any entity.
- TM2: Users must *only* trust the tag issuer.
- TM3: Users must trust the tag issuer *and* all readers.

To realize trust model TM1, the RFID scheme must provide full anonymity. However, this seems to be related to other systems as those for anonymous credentials and thus, TM1 seems to be possible only with high computational and communication resources, which is inappropriate for low-cost RFID devices (see Section 3.2).

In trust model TM2 users must at least trust the issuer with respect to their privacy whereas in trust model TM3 users must additionally trust the readers. Privacy to all entities outside the system (i.e., all unknown entities that are not trusted by the user) must be persevered in any case. Both trust models TM2 and TM3 can be achieved by existing RFID protocols. However, as discussed in Section 3, these protocols lack usability and thus are not applicable to most real-world scenarios.

It is assumed that all communication that takes place during the process of issuing a tag cannot be eavesdropped or manipulated by an adversary. This is reasonable in practice since a user may either use out-of-band communication or a secure channel to communicate to the tag issuer. However, following the traditional adversarial models, an adversary can eavesdrop all communication of a tag after it has been issued. Moreover, an adversary may perform active attacks on the corresponding protocols, which means that he can interact with all parties at the protocol level. Additionally, the adversary can corrupt (i.e., obtain the information stored on) tags. In trust model TM1 and TM2 the adversary is also allowed to corrupt the readers. In trust model TM1, the issuer may be compromised by an adversary who wants to violate privacy. In all other trust models the adversary cannot corrupt the issuer.

2.4   Requirement Analysis

We formally describe here the requirements of a dependable RFID system, where crucial security, privacy and usability properties have to be simultaneously achieved.

Location Privacy. Since RFID enables efficient detection and identification of a huge number of tags, a detailed dossier about user profiles (e.g., personal data and movements) can be created. The problem aggravates if tags can be associated with the identity of their corresponding users since this results in a complete loss of location privacy. Thus, to ensure location privacy, an RFID system must fulfill the following requirements:

- *Confidentiality*: Unauthorized access to user data should be infeasible.
- *Anonymity*: Unauthorized identification of tags should be infeasible.
- *Unlinkability*: Unauthorized tracing of tags should be infeasible.

Note that inexpensive RFID tags usually cannot provide expensive tamper-resistant hardware and thus, an adversary in practice can obtain the internal state of (i.e., all information stored on) a tag. Therefore, a stronger notion of location privacy is needed to capture traceability of tags in this case. To distinguish traceability in past or future protocol runs, [18] considers the notion of *forward* and *backward traceability*:

- *Backward traceability*: Accessing the current state of a tag should not allow to trace the tag in *previously recorded* protocol runs.
- *Forward traceability*: Accessing the current state of a tag should not allow to trace the tag in *future* protocol runs.

Security Goals. As mentioned in Section 2.2, one of the most important security goals is *authentication*. Thus no unauthorized user (i.e., who is not in possession of a legitimate tag) should be able to cheat any honest reader in order to obtain an unauthorized access. Another major requirement for any tag-based authentication scheme is the resilience to remote tampering with tags, which would allow denial-of-service attacks. We summarize the security goals as follows:

- *Authentication*: Only valid tags are accepted by honest readers.
- *Unclonability*: Duplication of valid tags must be infeasible.
- *Availability*: Unauthorized altering of tags must be infeasible.

In addition to the mentioned privacy and security goals it is important to consider some functional requirements desired for many real-world applications.

Functional Requirements. First, the manufacturing cost of a tag should be minimal, which means that the computational and storage requirements on a tag should be as low as possible. Additionally, verification of tags must be fast. For instance, it should be possible to verify a tag while a user is walking or shortly holding his tag near a reader (e.g., verification of an RFID e-ticket should be possible while entering a bus). Therefore, corresponding RFID protocols must be designed carefully to minimize the amount of computation and communication that must be performed without lowering the security and privacy requirements discussed above. Moreover, an RFID system should be able to handle a huge amount of tags. We summarize these goals as follows:

- *Efficiency*: Verification of tags must be fast.
- *Scalability*: A large amount of tags must be supported.

Depending on the underlying application scenario and the technological constraints, a practical realization may not be able to fulfill all of these goals and requirements. In particular, the security and functional requirements often contradict the privacy requirements.

2.5 Privacy Models for RFID

It is essential to carefully formalize the security and privacy goals discussed in Section 2.4 to enable the design of provably secure and privacy-preserving RFID protocols that are also usable in practice. Existing literature proposes various security and privacy models for RFID.

One of the first privacy definitions for RFID has been proposed by [28] and captures leakage of information on user-related data, including identities or movements of users. This definition is based on a security experiment where an adversary is challenged to distinguish a random value from the output of a legitimate tag. The definition of [28] also considers backwards traceability. However, this privacy model does not consider adversaries that can modify tags (e.g., by manipulating their memory) in order to trace them.

In [29], the author introduces a very restrictive adversary model specifically for RFID tags that cannot perform cryptographic operations. This model is based on assumptions on the number of queries an adversary can make to a tag, and aims at defining privacy to a broad range of real-world attacks. However, it does not allow the adversary to corrupt tags and thus does not capture forwards and backwards untraceability.

In [10] and [11], the author proposes a privacy model that provides various flexible definitions for different levels of privacy based on a security experiment where an adversary must distinguish two known tags. In [12], this model is extended by introducing the notion of *side-channel information* (i.e., whether authentication of a tag was successful or not). However, [12] does not capture backwards traceability since it does not allow an adversary to corrupt tags. In [13], the authors extend the definition of [12] by adding a *completeness* and *soundness* requirement, which means that a reader must accept *all* but *only* legitimate tags. The definition of [12] has been improved to consider backwards traceability [30].

Another approach to define privacy of RFID [8] is based on the universal composability (UC) framework [31]. This model claims to be the first that considers availability, which means that it captures security against denial-of-service attacks. However, it does not consider privacy of corrupted tags (i.e., backwards and forward traceability).

The author of [9] presents a privacy definition privacy that generalizes and classifies previous RFID privacy models by defining eight levels of privacy that correspond to different adversary models. The strongest adversary model in [9] covers all notions of privacy of previous schemes, including side-channel information, privacy of corrupted tags, and adversaries that can interact with tags and thus manipulate them at the protocol level. Moreover, the security definition of [9] is equivalent to [13]. In [32], the model of [9] has been extended to additionally consider authentication of readers to tags whereas [33] showed that the eight privacy classes of [9] can be reduced to three privacy classes under some restrictions on the power of the adversary.

3   Analysis of Existing Solutions

In this section we discuss in more detail previous works. Specifically, we will focus on stressing the main weaknesses of existing solutions with respect to the security, privacy and functional requirements discussed in Section 2.4.

3.1   Physical Methods

There is a body of literature that proposes physical solutions to enhance privacy of RFID. For instance, some RFID tags support a *kill command*, which is a tag-specific password programmed at manufacturing time that can be used to permanently disable a tag [34] such that it cannot be read any longer. This approach has been designed for electronic product labels that can be disabled after the corresponding product has left the supply chain and is given to the end-user. Another simple approach is to jam the RF interface of tags. The first solution is to put the tags into a Faraday cage, which is a container of metal mesh or foil that is opaque to RF signals (of a certain frequency). There are some vendors who already sell Faraday cages embedded into wallets, e.g., to protect RFID-enabled passports (e-pass) from unauthorized reading [35]. Alternatively, users may carry a special active jamming device that disturbs the radio signals of tags and readers in the user's vicinity [36].

Since all of these more or less radical approaches permanently disable the tags or require the user to interact with them, these solutions eliminate one of the main advantages of RFID. Thus, this chapter focuses on more sophisticated solutions that enhance user privacy by protocol-based techniques while maintaining the advantages of RFID.

3.2   Protocols for Anonymous Authentication

In an ideal RFID system, readers should learn nothing from the verification except that a tag is legitimate. It is possible to realize this by using privacy-preserving techniques like anonymous credential systems [37]. However, the use of anonymous credentials implies high computational (public-key cryptography) and typically also high communication (many rounds of interaction) requirements to all devices involved. Apparently, this does not comply to the functional requirements described in Section 2.4 and to the capabilities of current RFIDs [1, 4]. Thus, these techniques are not applicable unless powerful mobile computing devices (e.g., mobile phones or PDAs[2]) are used. However, the use of mobile computing devices has its own risks: These devices may run out of power (availability) and can be compromised by Trojans, which brings up new challenges (security), and many users do not yet own an RFID-enabled mobile phone that has sufficient computing power to run computationally demanding protocols like anonymous credential systems, e.g., as proposed in [37] (resource constraints).

3.3   Privacy-Preserving Protocols for RFID

There is a large body of literature on different approaches to implement privacy-preserving mechanisms for low-cost RFID tags. For instance, [7] gives a comprehensive overview of different solutions. The author classifies RFIDs as *basic tags* and *symmetric-key tags*. Basic tags refers to tags that have no computational and no cryptographic capabilities. Symmetric-key tags means tags that are capable of performing at least some symmetric cryptographic functions (e.g., random number generation, hashing, and encryption).

Protocols for Basic Tags. As basic tags cannot perform any cryptographic operation they disqualify for authentication purposes. Tags that only provide wireless readable memory can only forward the data stored in their memory and thus are subject to replay and cloning attacks. This means that all data stored on such a tag can be read and be used to create identical copies or to simulate the original tag to an honest reader. Another problem related to cloning is *swapping*. This means that an adversary can copy the data stored on tag A to another tag B and vice versa and thus change the identities of these tags. Therefore, basic tags cannot fulfill the requirements of authentication and unclonability.

Moreover, many solutions to enhance privacy of basic tags require tags to provide *many-writable* memory (e.g., [21, 22, 24]). The basic idea of these schemes is to frequently update the information stored on tags such that an adversary cannot link them. However, due to the lack of secure access control mechanisms it is impossible to prevent unauthorized writes to such tags. A simple denial-of-service attack is to write random data to a tag, which makes an honest reader

---

[2] An increasing number of mobile phones and PDAs supports the Near Field Communication (NFC) standard [38], which allows them to communicate to RFIDs.

to no longer accept the tag until it is reinitialized with correct data. This clearly violates the availability requirement. Moreover, an adversary could "mark" tags (e.g., store some recognizable data on them) such that he can track the tags even if they are frequently updated [24]. Obviously, this violates location privacy.

As a consequence, tags that provide no cryptographic functionality cannot be used in applications that require reliable authentication. Thus, it is inevitable to use tags that are capable of performing at least some cryptographic functions if authentication is of concern.

Protocols for Symmetric-Key Tags. A general problem of implementing privacy-preserving authentication based on symmetric-key cryptography is how to inform the other party which key must be used. Apparently, a tag cannot disclose its identity before the reader has been authenticated since this would violate unlinkability. Therefore, the reader does not know which authentication key it should use, and thus cannot authenticate to the tag. Essentially there are two approaches that address this problem: The first allows the reader to efficiently find the key used by the tag whereas the second frequently updates the identity of tags in a way that allows the reader to efficiently deduce the initial tag identity.

Key search approach. The basic idea of this approach has been introduced in [14]: Let $f_K(m)$ be a pseudorandom function on message m using key K. To authenticate to a reader, a tag first computes $h_T \leftarrow f_{K_T}(R)$ where K_T is a tag-specific key and R is a random value chosen by the tag. On receipt of (h_T, R), the reader forwards this tuple to a trusted server that computes $h_i \leftarrow f_{K_i}(R)$ for all keys $K_i \in \mathcal{K}$ where $\mathcal{K}$ denotes the set of the keys of all legitimate tags. The server accepts the tag if it finds a $K_i \in \mathcal{K}$ such that $h_i = h_T$. Finally, the server sends its decision whether to accept or to reject the tag to the reader. Since R is randomly chosen each time the tag is queried, the tag always emits a different tuple (h_T, R) which cannot be linked to the tuples sent in previous protocol runs. Moreover, the reader does not learn the identity (i.e., key K_T) of the tag since it only receives the response from the server. An obvious drawback of this solution is that the computational cost for the server to verify a tag is linear in the number of legitimate tags. Therefore, this basic approach does not fulfill the efficiency and scalability requirements. Another disadvantage of this solution is that all readers must have an online connection to the server, which, depending on the use case, may not be practical. Moreover, the tag must trust the server with respect to its privacy since the server can identify the tag when it finds the right key. Furthermore, this solution provides no security against replay-attacks (since an adversary may impersonate the tag by replaying any previously recorded tuple (h_T, R)) and thus violates authentication.

There are many subsequent works (including [2, 17, 18, 20]) that follow and optimize this approach by introducing new setup assumptions or by lowering the security or privacy requirements.

In [2], the authors improve the key search approach described above. The idea is to arrange the keys of all tags in a hierarchical tree. Each leaf of this tree corresponds to a tag, which means that all keys on the path from the root to a leaf are assigned to the corresponding tag. To authenticate to a reader, a tag runs one authentication protocol for each key it stores. Since all keys are arranged in a hierarchical tree, the reader must not search the whole key space. It is sufficient to search all keys of the first level of the subtree whose root is the key that has been used in the previous authentication protocol.

Assume the tree that stores all keys with depth d and branching degree b. Then this protocol can handle at most $n = b^d$ tags and each tag must store d keys. Moreover, verification of a tag requires the tag to run d authentication protocols with the reader. Compared to the basic approach of [14], the reader has to perform only $b \cdot d$ instead of $n = b^d$ computations to verify one single tag. However, since this scheme requires the tags to share several keys, compromise of one tag violates the location privacy of others [11]. Another drawback of this solution is that the number of tags in the system must be upper bounded before the scheme is initialized, which contradicts scalability.

The author of [19] improves the key search by various pre-computations during the creation of a tag. A tag is initialized with a key K_T, a counter state $t_i \leftarrow t_0$ and a maximum value t' for that counter. Moreover, the set of tuples $(t_0, f_{K_T}(t_0)), \ldots, (t', f_{K_i}(t'))$, where f is a pseudorandom function, is computed and added to the database of the reader. To authenticate a tag, the reader sends a value t_j to the tag. In case this value t_j has already been used (i.e., $t_j < t_i$) or exceeds t' (i.e., $t_j > t'$), the tag returns a random value h_j. Otherwise, the tag responds with $h_j \leftarrow f_{K_T}(t_j)$ and updates $t_i \leftarrow t_j$. The reader accepts the tag if it finds a tuple (t_j, h_j) in its database. According to [19], this protocol does not provide security against denial-of-service attacks, which violates availability. Moreover, [39] shows that this protocol does not provide unlinkability since it is possible to trace tags that have different maximum counter values. Clearly, this violates location privacy. Further, it does not provide authentication since an adversary may query a tag with different t_j and learn the corresponding responses h_j which later can be replayed to an honest reader [39].

The authors of [40] propose a scheme where the reader must only perform a few binary operations to test a key. The idea of this protocol is to blind a sufficiently large set of individual bits of the key K_T of the tag to be authenticated with random bits b_i such that $\sum K_T[i] \oplus b_i = C$, where C is some constant. To identify the tag, the reader recomputes this equation using all keys it knows. If it finds a match, it accepts the tag and rejects it otherwise. Unfortunately, after having recorded a large number of authentication protocol instances of the same tag, an adversary can reconstruct the key of the tag and thus is able to trace and to clone it [39]. This violates the untraceability and authentication requirements. Thus, this protocol does not provide location privacy.

Identity update approach. This approach relies on updating the identity of a tag each time it has been authenticated. Some of the protocols following this approach allow to authenticate a tag in constant time. However, these solutions require the readers to have permanent access to a trusted database that keeps track of the identity updates of all legitimate tags. As discussed above, this is inappropriate for many practical systems.

One of the first protocols following this approach has been presented in [16]. The authors propose a tag to update its state each time it is interrogated. Therefore, a tag is initialized with some initial identity T_0. Each time a reader communicates to the tag, the tag responds with $R_i \leftarrow g(K, T_i)$, where g is some one-way pseudorandom function and K is the authentication key of the tag. At the same time, the tag updates its identity to $T_{i+1} \leftarrow h(K, T_i)$ where h is a one-way pseudorandom function that is different from g. To identify the tag, the reader computes $R_i' \leftarrow g(K', h^i(K', T_0'))$ for all tuples (K', T_0') it knows until it finds an $R_i' = R_i$ for some $i \leq m$. This means that, as in the basic approach of [14], the verification of one single tag depends on the number of all legitimate tags in the system. Moreover, the maximum number of interrogations per tag is fixed to some value m. Thus, this protocol obviously does not fulfill the efficiency and scalability requirements. Moreover, an adversary may perform denial-of-service attacks since a tag can be invalidated by interrogating it more than m times, which clearly violates the availability requirement. However, this approach provides backwards traceability. Since h and g are one-way pseudorandom functions, an adversary who corrupted a tag cannot compute its preceding identities T_i nor can he recognize previous responses R_i of the tag.

The authors of [15] and [17] consider the problem of denial-of-service attacks and allow a tag to update its state only after the reader has been successfully authenticated to the tag. However, this allows tracing of tags between two successful authentications to a legitimate reader. The protocol proposed in [15] makes a tag to additionally transmit the number μ of interactions since the last successful authentication to a legitimate reader. This information is used by readers to speed up the identification of the tag and to prevent replay attacks. However, an adversary can trace tags by increasing the value μ to a very high value that he can recognize later [10]. Thus, this approach does not provide location privacy.

In [41], another protocol is proposed where each tag is assigned an authentication key K_T and identifier T. The reader has access to a database that contains a tuple (T, K_T) for each legitimate tag. To authenticate to a reader, the tag chooses some random number R and sends the blinded identifier $E \leftarrow R \cdot K_T + T$ to the reader. The reader accepts the tag if it finds a tuple (T, K_T) in his database for which $T = E \bmod K_T$. In the worst case, the reader must do this test for all tuples (T, K_T) of each tag it knows. Clearly, this violates the efficiency and scalability requirements.

Conclusion. As pointed out in this section, existing solutions for privacy-preserving RFID systems do either not provide location privacy (in the presence of real-world adversaries who can corrupt tags) or suffer from drawbacks like the possibility of denial-of-service or impersonation attacks as well as inefficient tag verification, which prevents their deployment in practice.

4   New Directions

In this section we explore some recent proposals for simplifying and improving the design of privacy-preserving RFID protocols.

4.1   Anonymizer-Based Protocols

An interesting approach to enhance privacy of RFID without lifting the computational requirements to tags are anonymizer-based protocols. These protocols rely on external devices (called *anonymizers*) that are in charge of providing anonymity of tags. Anonymizer-based RFID protocols are most suitable for many practical scenarios like electronic tickets and similar applications that must use a huge quantity of low-cost RFID tags with restricted capabilities.

The standard RFID system model can include anonymizers in different ways (see Figure 2). The most practical approach would be to integrate an anonymizer in each tag. However, due to computational constrains of current cost-efficient RFIDs this is not feasible. Another user-friendly approach is to provide public anonymizers. Since users of the RFID system must trust the anonymizers to frequently anonymize their tags and to prevent denial-of-service attacks by malicious anonymizers it is necessary that only authorized anonymizers can anonymize tags. In practice there may be a variety of public anonymizing service providers the user may choose (see Figure 2a). Alternatively, anonymizers may be controlled by the (trusted) operator of the RFID system (see Figure 2b). Anonymizers may be included into readers or mounted at the stations or in the vehicles of the transit enterprise. To further enhance the level of privacy, each user may be in possession of a personal anonymizer that can only be used to anonymize his own tags (see Figure 2c). Such a user-controlled anonymizer can be a dedicated hardware device (e.g., provided by the operator of the RFID system) or a software that runs on the mobile phone or PDA of the user. Note that in case a user's personal anonymizer runs out of power, the user will loose privacy until his anonymizer is operable again but he can still prove authorization using his RFID tags. Moreover, since there can be additional anonymizers in public places and their capabilities can be embedded into the readers (when this does not significantly affect the performance of the system), the user's location privacy is not completely lost.

Summing up, the user must trust the anonymizer with respect to privacy. However, this is a reasonable assumption since the anonymizer is either under the user's control or is managed by a trusted entity.

There are several proposals on privacy-preserving RFID protocols that employ anonymizers (e.g., [22, 24]). The main concept of these protocols is that an RFID tag stores a ciphertext that is sent to the reader each time it must authenticate itself. However, this ciphertext is static data that can be used to identify the tag and thus, must be frequently changed to provide anonymity of the tag. Due to the limited capabilities of currently available RFID technology [1, 4], tags are not capable of re-encrypting their ciphertexts on their own. Thus, privacy in these protocols relies on external anonymizers that frequently re-encrypt the ciphertexts.[3]

---

[3] Note that re-encryption does not always require knowledge of the keys that correspond to the ciphertext to be re-encrypted and thus, any entity may act as anonymizer.

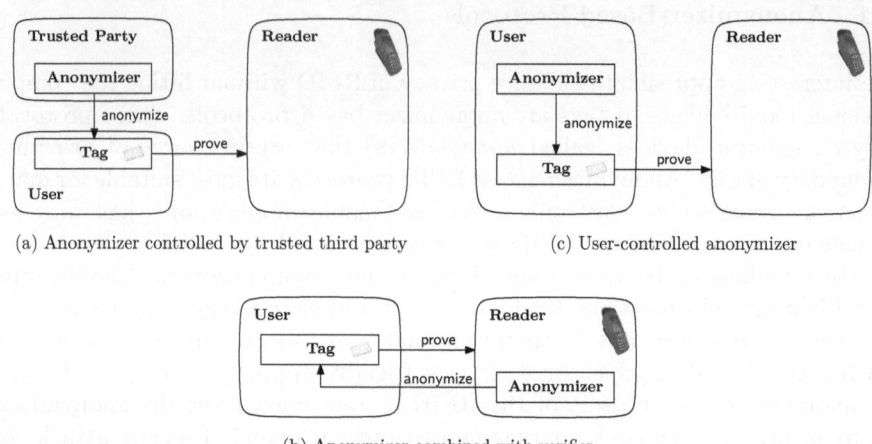

(a) Anonymizer controlled by trusted third party   (c) User-controlled anonymizer

(b) Anonymizer combined with verifier

Fig. 2. Integration of anonymizers into the RFID system model

The first proposal to use re-encryption was presented in [21] that considers a plan by the European Central Bank to embed RFID tags to Euro banknotes to aggravate forgeries [42]. The authors propose to store a ciphertext of the serial number of a banknote on the RFID tag that is attached to it. Each time the banknote is spent, anonymizers in shops or banks re-encrypt the ciphertext stored on the corresponding tag, which is supposed to render multiple appearances of a given RFID tag unlinkable. The drawback of this scheme is that the serial number of each banknote must be optically scanned before its ciphertext can be re-encrypted.

In [22], the authors introduce a primitive called *universal re-encryption*, that is an extension of the ElGamal encryption scheme where re-encryption is possible without knowledge of the corresponding (private and public) keys [24]. A drawback of this approach is that an adversary can "mark" tags such that he can recognize them even after they have been re-randomized. This problem has been first addressed in [23], which shows tracing attacks and also proposes solutions to solve this issue. In [24], the authors improve the ideas of [22] and [23] by introducing the notion of *insubvertible encryption*, which adds a signature on the blinded public-key of the reader that is linked to the ciphertexts that are stored on the tags. Re-randomization involves this signature in a way that prevents an adversary from marking tags.

However, all of these schemes do not provide security against denial-of-service attacks since any entity that impersonates an anonymizer can permanently disable tags by writing random data to them. This issue has been addressed in [25] where the authors suggest to introduce authentication of anonymizers to tags. They propose to use re-randomizable encryption to distribute the symmetric authentication key associated with a tag only to authorized (i.e., trusted) readers and anonymizers, which ensures unlinkability to any unauthorized entity outside the RFID system.

Moreover, existing security and privacy models (see Section 2.5) cannot be directly applied to RFID systems that are based on anonymizers. However, anonymizers can be very useful tools to achieve privacy with cost-efficient tags. Thus, a security and privacy model that also considers such additional players is of interest for the design of secure and privacy-preserving anonymizer-based RFID systems.

4.2   Physically Unclonable Functions (PUFs)

To prevent cloning of a tag it must be infeasible to determine its authentication secrets by both attacking the corresponding authentication protocols as well as by physically attacking the tag. One solution to counterfeit cloning attacks is to employ physical protection mechanisms that aggravate reading out the memory of a tag [43, 44]. However, this would dramatically increase the price of tags and render them inappropriate for most commercial applications. A more economic solution to prevent cloning can be implemented by using physically unclonable functions (PUFs) [27, 45].

A PUF consists of an inherently unclonable noisy function Π that is embedded into a physical object [46]. The unclonability of a PUF comes from randomness generated during its manufacturing processes. A PUF maps challenges to responses. A *challenge* c is a stimulus signal input to the PUF on which the PUF returns a *response* $r' = \Pi(c)$ that is specific for that PUF w.r.t. to the stimulus c. This response r' relies on physical properties of the corresponding physical object, which is subject to environmental noise (e.g., temperature or voltage). Thus, the physical component of a PUF will always return slightly different responses r' to the same stimulus c. These slight deviations can be removed by a small circuit, called *fuzzy extractor*, that (up to a certain threshold) maps different responses r' to a unique value r for each specific challenge c. The fuzzy extractor needs some additional input w (called *helper data*) to remove the effects of noise on the physical component Π.

Two different PUFs that are challenged with the same stimulus will return different responses with overwhelming probability. Moreover, a PUF can be embedded into a microchip, e.g., by exploiting statistical variations of delays of gates and wires within the chip [45]. These deviations are unique for every sample from a set of chips (even from the same lot or wafer) that implement the same circuit.

One of the first proposals on using PUFs for RFID is introduced by [47]. It proposes the manufacturer of a tag to store a set of challenge-response pairs in a database, which can later be used by RFID readers that are connected to this database to identify a tag. The idea is that the reader chooses a challenge from the database, queries the tag and checks whether the database contains a tuple that matches the response received from the tag. One problem of this approach is that a challenge-response pair cannot be reused since this would enable replay attacks and allow to trace tags. This scheme has been implemented by [48] who provide a realization of PUF-enabled RFID tags and analyze their security and usability.

The authors of [49] propose a similar approach based on the physical characteristics of SRAM cells. The advantage of this approach is that SRAM-PUFs can be implemented using the existing SRAM memory cells of the RFID chip without the need for additional hardware.

In [27], the authors propose to use a PUF as secure key storage for the secret authentication key of an RFID tag. This means that instead of storing the key in some protected memory, a PUF is used to reconstruct the key whenever it is needed. Since the key is inherently hidden within the physical structure of the PUF, obtaining this secret by hardware-related attacks is supposed to be intractable for real-world adversaries [45]. According to [27], a PUF-based key storage can be implemented with less than 1000 gates. However, their authentication scheme relies on public-key cryptography, which is still much too expansive for current commercial RFID tags.

The authors of [50] follow the approach of frequently updating the identity of tags to provide privacy (see Section 3.3) and suggest to use PUFs instead of hash functions. They propose to equip each tag T with a PUF that is used to derive new tag identifiers. Since readers cannot recompute these identifiers, the authors propose them to access a database that stores a tuple $(T, \mathsf{PUF}(T), \mathsf{PUF}^2(T), \ldots, \mathsf{PUF}^m(T))$ for each legitimate tag T where $\mathsf{PUF}^k(T) = r_k$ is computed as $r_{i+1} = \mathsf{PUF}(r_i)$ for $i \in [1, k]$ and $r_0 = T$.

To authenticate to a reader, a tag first sends its current identifier T_i and then updates its identity to $T_{i+1} \leftarrow \mathsf{PUF}(T_i)$. The reader then checks whether there is a tuple that contains a value T_i in the database. In case the reader finds T_i, it accepts the tag and invalidates all previous database entries for that tag to prevent replay attacks. A major drawback of this scheme is that a tag can only authenticate m times without being re-initialized, which, as the authors mention, allows an adversary to perform denial-of-service attacks.

Physically unclonable functions are a very interesting and promising approach to increase the security of existing RFID systems. Moreover, they open new directions towards cost-efficient privacy-preserving protocols based on physical assumptions. They provide cost-effective and practical tamper-evident storage for cryptographic secrets that even cannot be learned or reproduced by the manufacturer of the corresponding PUF.

However, several aspects of PUFs and their deployment to RFID require further research. Since PUFs are bound to the device in which they are embedded, no other entity can verify the response r of a PUF to a given challenge c without knowing an authentic tuple (c, r) in advance. As we have already mentioned, current PUF-based RFID protocols aim at circumventing this problem by providing the reader with a database that contains a set of challenge-response pairs that act as reference values for the responses of the interrogated PUF. However, this approach opens the possibility for denial-of-service and replay-attacks. Another problem with PUFs is that their realizations require careful statistical testing before they can be safely deployed to real security-critical products. Moreover, to our knowledge, there is no complete security and adversary model for PUFs yet.

4.3 Public-Key Cryptography

In the near future, it is hoped that RFID tags capable of performing public key cryptography are available [4]. As shown in [9] public-key cryptography can be used to design (stateless) protocols that provide location privacy (including backward and forward untraceability).

The public-key-based protocol proposed in [9] is a simple challenge-response protocol. Each tag is initialized with an identifier T, a tag-specific authentication secret K_T and the public key pk of the reader. The reader knows the corresponding secret key sk, a master secret K' and a function f such that $f_{K'}(T) = K_T$. To verify the authenticity of a tag, the reader challenges the tag with some random value R. The tag responds with $C \leftarrow \mathsf{Enc}_{pk}(T, K_T, R)$. To verify the authenticity of the tag, the reader first decrypts C and checks if the resulting plaintext contains the random value R that has been previously sent to the tag and if $K_T = f_{K'}(T)$. In case all checks pass the reader accepts the tag.

Recent literature presents several proof-of-concept implementations of public-key cryptography on passive RFID tags (i.e., those that are powered by the RFID readers) [51, 52].

5 Use Case: E-Tickets

Electronic tickets for public transportation is one of many practical RFID-based applications that is already widely deployed in practice and will become more popular in future [3, 5, 53]. However, RFID e-ticket systems currently used in practice do usually not consider privacy aspects (i.e., the confidentiality of the identity and location of users). Moreover, e-tickets must fulfill strict usability requirements in order to be competitive to conventional paper-based tickets.

5.1 General Scenario

An e-ticket system, as shown in Figure 3, consists of at least one ticket issuing entity (*issuer*), a set of *users*, *tickets*, and *readers*, who verify whether tickets are valid. Since we are focusing on RFID-based systems where tickets are realized as RFID tags, in the following we use *ticket* synonymously to *tag*.

Typically, a user U must obtain a ticket T from an issuer I. Therefore, user U selects his desired ticket. Issuer I then checks whether user U is eligible to obtain that ticket (e.g., whether U paid for it), and, if applicable, issues the ticket T and passes it to U. From now on, user U is able to use ticket T to prove that he is authorized to use the transit network. This means that every user who is in possession of a ticket that has been issued by a genuine issuer is considered to be an *authorized user*.

Now assume that, as shown in Figure 3, user U wants to travel from a place X to some location Y. Before U is allowed to enter the transit system at X, he must first prove to a reader V_{in} at the entrance of the transit network that he is authorized to access it. If reader V_{in} can successfully verify the user's ticket, user U is allowed to enter. Otherwise access will be denied. During his trip,

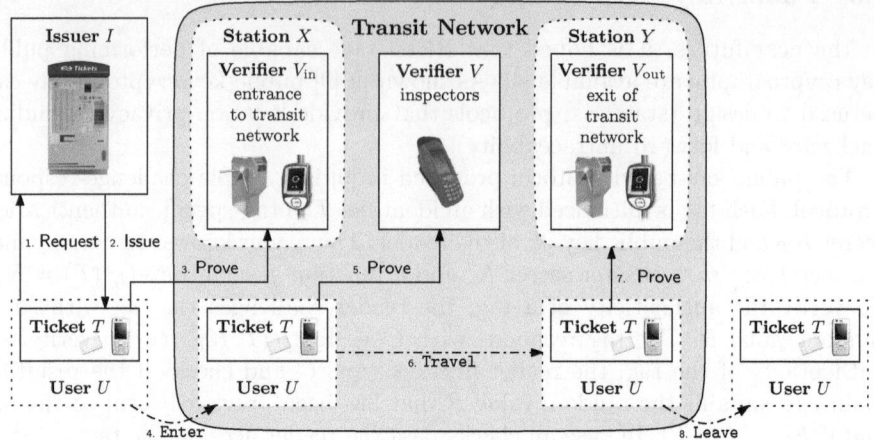

Fig. 3. General scenario of e-tickets

user U may encounter arbitrary inspections where he must prove that he is authorized to use the transit network. Thus, a reader V may check the user's ticket T. If verification of the ticket T is successful, user U is allowed to continue his trip. Otherwise, user U must leave the transit network and may be punished for using it without authorization. After arriving at Y, the user's ticket T may be checked for a last time. Again, if the ticket T cannot be verified successfully, user U may be punished.

Note that authentication is typically bound to some limitations. For instance, this may be some geographical or timely usage restriction that must also be considered during ticket verification.

5.2 E-Ticket Systems in Practice

Most e-ticket systems in practice are proprietary solutions whose specifications are not publicly available. This section reviews the most common approach of implementing authentication of e-tickets in practice by the Calypso e-ticket system [3, 54], of which at least some information is public. Moreover, to the best of our knowledge, there is no solution for RFID-based e-tickets in practice that explicitly considers the location privacy of users.

Calypso E-Ticket Standard. Calypso is an e-ticket standard based on RFID that is widely used in Europe and North and South America [3]. The roles in the Calypso system correspond to the model presented in Figure 3. However, Calypso does not consider privacy of users and thus does not fulfill any of the privacy requirements of Section 2.4. Actually, all transactions involving a Calypso e-ticket provide no confidentiality at all [54]. Moreover, Calypso tickets can store personal data of their owner (e.g., his name) that can be queried by every verifier. Thus the Calypso e-ticket system leaks user-related information and allows the creation of location profiles by everyone who is in possession of a standard RFID reader. All messages of a Calypso ticket are authenticated by a symmetric-key-based authentication mechanism. Calypso seems to fulfill all of the security requirements but none of the privacy requirements of Section 2.4.

Calypso implements a common approach to authenticate low-cost RFID tags based on a simple challenge-response protocol. Each tag has a symmetric authentication key K_T that can be computed as a function of the serial number S_T of the tag and a global master secret. All readers are equipped with a tamper-resistant security module that knows and protects this master secret and can be used as a black-box to compute K_T from S_T. To authenticate a tag, a reader sends a random challenge N_V to the tag, which then computes $H_T \leftarrow f(K_T, N_V)$ where f is some pseudorandom function. Finally, the tag returns (S_T, H_T) to the reader that uses its security module to derive K_T and then verifies H_T. If verification is successful, the tag has been authenticated. Obviously, this approach cannot provide location privacy since all transactions of a tag can be linked by its serial number S_T that is transmitted in clear in every protocol run. All subsequent transactions to update or to read data from a Calypso ticket are authenticated this way but are not encrypted.

Other E-Ticket Systems. There are several other proprietary solutions for e-tickets in practice. Most of them are based on widely used RFID tags. Prominent examples are FeliCa [5] and MiFare [53]. FeliCa [5] is a contactless smartcard sold by Sony that is mainly used in the Asia-Pacific area for different purposes, including e-tickets for public transportation. MiFare is a family of contactless smartcards produced by Philips/NXP Semiconductors. These tags are widely used for different purposes, including e-tickets for public transportation. There were several publications on attacks against MiFare Classic tags [55, 56, 57], that use a proprietary encryption algorithm which has been completely broken [58]. However, other MiFare products are claimed not to be affected.

The attacks on MiFare Classic tags demonstrate a major problem of proprietary security solutions: Manufacturers of low-cost hardware try to find a compromise between efficiency and security of their products. Thus, they often implement proprietary lightweight crypto algorithms whose specifications are not public, and thus are typically not sufficiently evaluated. As for MiFare Classic, these algorithms can often be reverse-engineered, which allows cryptanalysis or efficient key search by running the algorithms on more powerful hardware. In case of MiFare Classic, both ways allowed to break the security goals of these tags at a point in time where they were already widely used in practice.

5.3   Privacy-Preserving Protocols for E-Tickets

Privacy-preserving e-tickets are discussed in a few papers. In [37], the authors sketch an anonymous payment system for public transit based on anonymous credentials [59] and e-cash [60]. They propose tickets to be managed either by RFID tags or mobile computing devices like mobile phones or PDAs. As pointed out in Section 3.2, anonymous credentials and e-cash are not applicable to currently available RFID devices whereas the use of mobile phones or PDAs for managing e-tickets introduces several other drawbacks.

In [25], the authors describe a framework for anonymizer-based protocols for RFID e-tickets (see Section 4.1). This framework requires tags to provide a random number generator, symmetric-key cryptography (i.e., keyed hashing) and a key storage based on physically unclonable functions, which are realistic assumptions for current RFID technology [1, 4, 27, 48]. Authentication relies on a two-pass authentication protocol using symmetric cryptography only. The corresponding authentication keys are stored in a secure PUF-based key storage on the tag (see Section 4.2). To allow readers to identify a tag without revealing the tag identity to unauthorized third parties, re-randomizable public-key encryption [22, 24] is used. Therefore, each tag stores a re-randomizable ciphertext of its symmetric authentication key. This ciphertext is passed to the reader each time the tag is interrogated. Only authorized readers can decrypt this ciphertext and obtain the symmetric authentication key that acts as identifier for the tag and that is used in the symmetric authentication protocol with the tag. Additionally, the ciphertext contains a signature on the authentication key that has been issued by the tag issuing entity. This signature can be used by the reader to verify the authenticity of the tag. To prevent denial-of-service attacks, anonymizers must authenticate to tags. This works similarly to the authentication of tags to readers.

6   Open Problems

We identify the following open problems: The design of provably-secure privacy-preserving protocols that are applicable to real-world scenarios is very challenging. According to [9], it is an open problem whether location privacy can actually be achieved by using RFID tags that are not capable of performing public-key cryptography.

Another problem concerns different ad-hoc notions of RFID security and privacy that are often incomparable to each other [30]. Consequently, there are several protocols that can be proven secure in one privacy model but are insecure in other models. For instance, the OSK protocol [28] can be proven secure in the model of [10] although a tracing attack can be shown in the model of [12]. Therefore, it is crucial to define a sufficiently general privacy model for RFID in order to design secure privacy-preserving RFID protocols.

There are also other, more non-technical issues, which concern the user-awareness and education regarding privacy. Also time to market aspects force companies to leave out privacy aspects in their products unless they are required by law or by consumer protection organizations.

References

1. Atmel Corporation: Innovative IDIC solutions (2007),
   `http://www.atmel.com/dyn/resources/prod_documents/doc4602.pdf`
2. Molnar, D., Wagner, D.: Privacy and security in library RFID: Issues, practices, and architectures. In: Proceedings of the 11th ACM Conference on Computer and Communications Security, pp. 210–219. ACM Press, New York (2004)
3. Calypso Networks Association: Web site of Calypso Networks Association. (May 2007), `http://www.calypsonet-asso.org/`
4. NXP Semiconductors: MIFARE smartcard ICs. (September 2008),
   `http://www.mifare.net/products/smartcardics/`
5. Sony Global: Web site of Sony FeliCa. (June 2008),
   `http://www.sony.net/Products/felica/`
6. I.C.A. Organization: Machine Readable Travel Documents, Doc. 9303, Part 1 Machine Readable Passports, 5th edn (2003)
7. Juels, A.: RFID security and privacy: A research survey. Journal of Selected Areas in Communication 24(2), 381–395 (2006)
8. Burmester, M., van Le, T., de Medeiros, B.: Provably secure ubiquitous systems: Universally composable RFID authentication protocols. In: Proceedings of Second International Conference on Security and Privacy in Communication Networks (SecureComm), pp. 1–9. IEEE Computer Society, Los Alamitos (2006)
9. Vaudenay, S.: On privacy models for RFID. In: Kurosawa, K. (ed.) ASIACRYPT 2007. LNCS, vol. 4833, pp. 68–87. Springer, Heidelberg (2007)
10. Avoine, G.: Adversarial model for radio frequency identification. Cryptology ePrint Archive, Report 2005/049 (2005)
11. Avoine, G., Dysli, E., Oechslin, P.: Reducing time complexity in RFID systems. In: Preneel, B., Tavares, S. (eds.) SAC 2005. LNCS, vol. 3897, pp. 291–306. Springer, Heidelberg (2006)
12. Juels, A., Weis, S.A.: Defining strong privacy for RFID. Cryptology ePrint Archive, Report 2006/137 (2006)
13. Damgård, I., Østergaard, M.: RFID security: Tradeoffs between security and efficiency. In: RSA Conference, Cryptographers' Track, pp. 318–332 (2008)
14. Weis, S.A., Sarma, S.E., Rivest, R.L., Engels, D.W.: Security and privacy aspects of low-cost radio frequency identification systems. In: Hutter, D., Müller, G., Stephan, W., Ullmann, M. (eds.) Security in Pervasive Computing. LNCS, vol. 2802, pp. 201–212. Springer, Heidelberg (2004)
15. Henrici, D., Müller, P.: Hash-based enhancement of location privacy for radio-frequency identification devices using varying identifiers. In: Proceedings of the Second IEEE Annual Conference on Pervasive Computing and Communications Workshops, pp. 149–153. IEEE Computer Society, Los Alamitos (2004)
16. Ohkubo, M., Suzuki, K., Kinoshita, S.: Efficient hash-chain based RFID privacy protection scheme. In: International Conference on Ubiquitous Computing (UbiComp), Workshop Privacy: Current Status and Future Directions (September 2004)
17. Dimitriou, T.: A lightweight RFID protocol to protect against traceability and cloning attacks. In: Proceedings of the First International Conference on Security and Privacy for Emerging Areas in Communications Networks (SecureComm), pp. 59–66. IEEE Computer Society, Los Alamitos (2005)
18. Lim, C.H., Kwon, T.: Strong and robust RFID authentication enabling perfect ownership transfer. In: Ning, P., Qing, S., Li, N. (eds.) ICICS 2006. LNCS, vol. 4307, pp. 1–20. Springer, Heidelberg (2006)

19. Tsudik, G.: YA-TRAP: Yet Another Trivial RFID Authentication Protocol. In: Security in Pervasive Computing. LNCS, vol. 2802, pp. 640–643. IEEE Computer Society, Los Alamitos (2006)
20. Song, B., Mitchell, C.J.: RFID authentication protocol for low-cost tags. In: Proceedings of the First ACM Conference on Wireless Network Security, pp. 140–147. ACM Press, New York (2008)
21. Juels, A., Pappu, R.: Squealing Euros: Privacy protection in RFID-enabled banknotes. In: Wright, R.N. (ed.) FC 2003. LNCS, vol. 2742, pp. 103–121. Springer, Heidelberg (2003)
22. Golle, P., Jakobsson, M., Juels, A., Syverson, P.: Universal re-encryption for mixnets. In: Okamoto, T. (ed.) CT-RSA 2004. LNCS, vol. 2964, pp. 163–178. Springer, Heidelberg (2004)
23. Saito, J., Ryou, J.C., Sakurai, K.: Enhancing privacy of universal re-encryption scheme for RFID tags. In: Yang, L.T., Guo, M., Gao, G.R., Jha, N.K. (eds.) EUC 2004. LNCS, vol. 3207, pp. 879–890. Springer, Heidelberg (2004)
24. Ateniese, G., Camenisch, J., de Medeiros, B.: Untraceable RFID tags via insubvertible encryption. In: Proceedings of the 12th ACM Conference on Computer and Communications Security, pp. 92–101. ACM Press, New York (2005)
25. Sadeghi, A.R., Visconti, I., Wachsmann, C.: User privacy in transport systems based on RFID e-tickets. In: International Workshop on Privacy in Location-Based Applications (PiLBA), Malaga, Spain (October 9, 2008)
26. NXP Semiconductors: MIFARE application directory (MAD) — list of registered applications. (April 2008), http://www.nxp.com/acrobat/other/identification/mad_overview_042008.pdf
27. Tuyls, P., Batina, L.: RFID-tags for anti-counterfeiting. In: Pointcheval, D. (ed.) CT-RSA 2006. LNCS, vol. 3860, pp. 115–131. Springer, Heidelberg (2006)
28. Ohkubo, M., Suzuki, K., Kinoshita, S.: Cryptographic approach to "privacy-friendly" tags (November 2003)
29. Juels, A.: Minimalist cryptography for low-cost RFID tags (extended abstract). In: Blundo, C., Cimato, S. (eds.) SCN 2004. LNCS, vol. 3352, pp. 149–164. Springer, Heidelberg (2005)
30. Ha, J.H., Moon, S.J., Zhou, J., Ha, J.C.: A new formal proof model for RFID location privacy, In: [61], pp. 267–281
31. Canetti, R.: Universally Composable Security: a New Paradigm for Cryptographic Protocols. In: 42nd Symposium on Foundations of Computer Science (FOCS 2001), 1109 Spring Street, Suite 300, Silver Spring, MD 20910, USA, pp. 136–145. IEEE Computer Society Press, Los Alamitos (2001)
32. Paise, R.I., Vaudenay, S.: Mutual authentication in RFID: Security and privacy. In: ASIACCS 2008: Proceedings of the 2008 ACM Symposium on Information, Computer and Communications Security, pp. 292–299. ACM Press, New York (2008)
33. Ng, C.Y., Susilo, W., Mu, Y., Safavi-Naini, R.: RFID privacy models revisited, In: [61], pp. 251–256
34. EPCglobal Inc.: Specification for RFID air interface — EPC radio-frequency protocols, Class-1 Generation-2 UHF RFID, protocol for communications at 860 MHz–960 MHz, version 1.1.0 (December 2005)
35. DIFRwear: Web site of difrwear (January 2009), http://www.difrwear.com/products.shtml
36. Peris-Lopez, P., Hernandez-Castro, J.C., Estevez-Tapiador, J.M., Ribagorda, A.: RFID systems: A survey on security threats and proposed solutions. In: Cuenca, P., Orozco-Barbosa, L. (eds.) PWC 2006. LNCS, vol. 4217, pp. 159–170. Springer, Heidelberg (2006)

37. Heydt-Benjamin, T.S., Chae, H.J., Defend, B., Fu, K.: Privacy for public transportation. In: Danezis, G., Golle, P. (eds.) PET 2006. LNCS, vol. 4258, pp. 1–19. Springer, Heidelberg (2006)
38. NFC Forum: Web site of Near Field Communication (NFC) Forum (April 2008), http://www.nfc-forum.org/
39. Ouafi, K., Phan, R.C.W.: Privacy of recent RFID authentication protocols. In: Chen, L., Mu, Y., Susilo, W. (eds.) ISPEC 2008. LNCS, vol. 4991, pp. 263–277. Springer, Heidelberg (2008)
40. Castelluccia, C., Soos, M.: Secret shuffling: A novel approach to RFID private identification. In: Conference on RFID Security 2007, Malaga, Spain (July 11–13, 2007)
41. Mitra, M.: Privacy for rfid systems to prevent tracking and cloning. International Journal of Computer Science and Network Security 8(1), 1–5 (2008)
42. Economist: Security technology: Where's the smart money? The Economist, 69–70 (February 2002)
43. Skorobogatov, S.P., Anderson, R.J.: Optical fault induction attacks. In: 4th International Workshop on Cryptographic Hardware and Embedded Systems (CHES 2002), Redwood Shores, CA, USA, August 13–15, 2002, Revised Papers. Volume 2523 of LNCS. Springer Verlag (2002) 31–48
44. Neve, M., Peeters, E., Samyde, D., Quisquater, J.J.: Memories: A survey of their secure uses in smart cards. In: Proceedings of the Second IEEE International Security in Storage Workshop, October 31, 2003, pp. 62–72. IEEE Computer Society, Los Alamitos (2003)
45. Gassend, B., Clarke, D., van Dijk, M., Devadas, S.: In: Proceedings of the 18th Annual Computer Security Applications Conference, December 9–13, 2002, pp. 149–160. IEEE Computer Society, Los Alamitos (2002)
46. Tuyls, P., Škoriç, B., Kevenaar, T. (eds.): Security with Noisy Data — On Private Biometrics, Secure Key Storage, and Anti-Counterfeiting. Springer, Heidelberg (2007)
47. Ranasinghe, D.C., Engels, D.W., Cole, P.H.: Security and privacy: Modest proposals for low-cost rfid systems. In: Auto-ID Labs Research Workshop (September 2004)
48. Devadas, S., Suh, E., Paral, S., Sowell, R., Ziola, T., Khandelwal, V.: Design and implementation of PUF-based unclonable RFID ICs for anti-counterfeiting and security applications. In: IEEE International Conference on RFID 2008, April 16-17, pp. 58–64. IEEE Computer Society, Las Vegas (2008)
49. Holcomb, D.E., Burleson, W.P., Fu, K.: Initial SRAM state as a fingerprint and source of true random numbers for RFID tags. In: Conference on RFID Security 2007, Malaga, Spain (July 11–13, 2007)
50. Bolotnyy, L., Robins, G.: Physically unclonable function-based security and privacy in RFID systems. In: Proceedings of the Fifth IEEE International Conference on Pervasive Computing and Communications, pp. 211–220. IEEE Computer Society, Los Alamitos (2007)
51. Hein, D., Wolkerstorfer, J., Felber, N.: ECC is ready for RFID — a proof in silicon. In: Conference on RFID Security 2007, Malaga, Spain (July 11-13, 2007)
52. Oren, Y., Feldhofer, M.: WIPR — a public key implementation on two grains of sand. In: Conference on RFID Security 2007, Malaga, Spain (July 11-13, 2007)
53. NXP Semiconductors: Web site of MIFARE (May 2007), http://mifare.net/
54. Spirtech: CALYPSO functional specification: Card application, version 1.3. (October 2005), http://calypso.spirtech.net/

55. Nohl, K., Plötz, H.: MiFare — Little security despite obscurity (2007),
    http://events.ccc.de/congress/2007/Fahrplan/events/2378.en.html
56. Schreur, R.W., van Rossum, P., Garcia, F., Teepe, W., Hoepman, J.H., Jacobs, B.,
    de Koning Gans, G., Verdult, R., Muijrers, R., Kali, R., Kali, V.: Security flaw in
    MiFare Classic (March 2008),
    http://www.sos.cs.ru.nl/applications/rfid/pressrelease.en.html
57. Garcia, F.D., de Koning Gans, G., Muijrers, R., van Rossum, P., Verdult, R.,
    Schreur, R.W., Jacobs, B.: Dismantling mifare classic. In: [61], pp. 97–114
58. Courtois, N.T., Nohl, K., O'Neil, S.: Algebraic attacks on the Crypto-1 stream
    cipher in MiFare Classic and Oyster Cards. Cryptology ePrint Archive, Report
    2008/166 (2008)
59. Camenisch, J., Lysyanskaya, A.: Signature schemes and anonymous credentials
    from bilinear maps. In: Franklin, M. (ed.) CRYPTO 2004. LNCS, vol. 3152, pp.
    56–72. Springer, Heidelberg (2004)
60. Camenisch, J., Hohenberger, S., Lysyanskaya, A.: Compact e-cash. In: Cramer,
    R. (ed.) EUROCRYPT 2005. LNCS, vol. 3494, pp. 302–321. Springer, Heidelberg
    (2005)
61. Jajodia, S., Lopez, J. (eds.): Computer Security — ESORICS 2008. LNCS,
    vol. 5283. Springer, Heidelberg (2008)

Privacy in Georeferenced Context-Aware Services: A Survey

Daniele Riboni, Linda Pareschi, and Claudio Bettini

EveryWare Lab, D.I.Co., University of Milano
via Comelico 39, I-20135 Milano, Italy
{riboni,pareschi,bettini}@dico.unimi.it

Abstract. Location based services (LBS) are a specific instance of a broader class of Internet services that are predicted to become popular in a near future: context-aware services. The privacy concerns that LBS have raised are likely to become even more serious when several context data, other than location and time, are sent to service providers as part of an Internet request. This paper provides a classification and a brief survey of the privacy preservation techniques that have been proposed for this type of services. After identifying the benefits and shortcomings of each class of techniques, the paper proposes a combined approach to achieve a more comprehensive solution for privacy preservation in georeferenced context-aware services.

1 Introduction

It is widely recognized that the success of context-aware services is conditioned to the availability of effective privacy protection mechanisms (e.g., [1,2]). Techniques for privacy protection have been thoroughly studied in the field of databases, in order to protect microdata released from large repositories. Recently some of these techniques have been extended and integrated with new ones to preserve the privacy of users of Location Based Services (LBS) against possibly untrusted service providers as well as against other types of adversaries [3]. The domain of service provisioning based on location and time of request introduces novel challenges with respect to traditional privacy protection in microdata release. This is mainly due to the dynamic nature of the service paradigm, which requires a form of *online* privacy preservation technique as opposed to an *offline* one used, for example, in the publication of a view from a database. In the case of LBS, specific techniques are also necessary to process the spatio-temporal information describing location and time of request, which is also very dynamic. On the other hand, location and time are only two of the possibly many parameters characterizing the context of an Internet service request. Indeed, context information goes far beyond location and time, including data such as personal preferences and interests, current activity, physiological and emotional status, and data collected from body-worn or environmental sensors, just to name a few. As shown by Riboni et al. in [4], privacy protection techniques specifically developed for LBS are often insufficient and/or inadequate when applied to generic context-aware services.

C. Bettini et al. (Eds.): Privacy in Location-Based Applications, LNCS 5599, pp. 151–172, 2009.

Consider, for instance, cryptographic techniques proposed for LBS (e.g., [5,6]). These techniques provide strong privacy guarantees at the cost of high computational overhead on both the client and server side; moreover, they introduce expensive communication costs. Hence, while they may be profitably applied to simple LBS such as nearest neighbor services, it is unlikely that they would be practical for complex context-aware services. On the other hand, obfuscation techniques proposed for LBS (e.g., [7,8]) are specifically addressed to location information; hence, those techniques cannot be straightforwardly applied to other contextual domains. With respect to techniques based on identity anonymity in LBS (e.g., [9,10]) we point out that, since many other kinds of context data besides location may help an adversary in identifying the owner of those data, the amount of context data to be generalized in order to enforce anonymity is large. Hence, even if filtering techniques can be used for improving the service response, as shown by Aggarwal in [11], it could happen that in order to achieve the desired anonymity level, context data become too general to provide the service at an acceptable quality level. For this reason, specific anonymity techniques for generic context-aware services are needed.

Moreover, in pervasive computing environments context-aware services can exploit data provided by sensors deployed in the environment that can constantly monitor context data. Hence, if those context sources are compromised, an adversary's inference abilities may increase by taking advantage of the observation of users' behavior, and by knowledge of context information about those users. Defense techniques for privacy preservation proposed for LBS do not consider this kind of inference capabilities, since location and time are the only contextual parameters that are taken into account. As a result, protecting against the above mentioned kind of attacks requires not only novel techniques, but also different benchmarking tools for testing the efficiency and effectiveness of defense techniques. Fulfilling the latter requirement is particularly challenging; indeed, while in LBS several efforts have been made in order to collect real location data to be used for benchmarking, gathering a wide set of context data for the same purpose is even more difficult due to both technical difficulties in gathering those data, and users' reluctance in disclosing potentially sensitive information. In order to address this issue, one of the most common method consists in developing ad-hoc simulations of context-aware services and scenarios. These simulations are based on statistical analysis of real environments and on the generation of synthetic context data. However, we point out that modeling realistic context-aware scenarios is particularly difficult, since those scenarios are characterized by a variety of possible contextual conditions, which in turn influence service responses and users' behaviors.

As regards users' privacy requirements, we claim that context-awareness emphasizes the need for personalized privacy preferences. Indeed, users' privacy risk perception is strongly affected by personal experiences and context, and it may significantly vary from an individual to another. Hence, it becomes fundamental for users issuing requests for context-aware services to have the possibility of

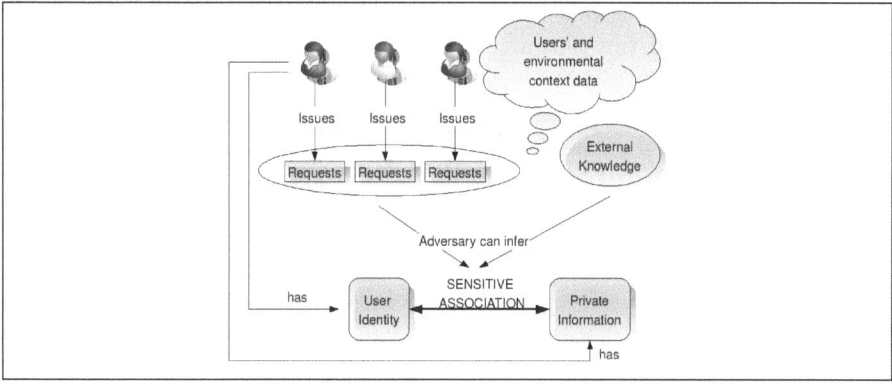

Fig. 1. The privacy threat

setting detailed privacy preferences with respect to the service they are asking for, the sensitive data involved in service adaptation, and the contextual situation.

As depicted in Figure 1, the general privacy threat we are facing is the release of *sensitive associations* between a user's identity and the information that she considers private. The actual privacy risk certainly depends on the adversary's model; for the purpose of this survey, unless we mention specific attacks, we adopt the general assumption that an adversary may obtain service requests and responses as well as publicly available information.

We distinguish different types of defense techniques that can be used to contrast the privacy threat.

○ **Network and cryptographic protocols.** These are mainly used to avoid that an adversary can access the content of a request or response while it is transmitted as well as to avoid that a network address identifies the location and/or the issuer of a request.

○ **Access control mechanisms.** These are used to discriminate (possibly based on context itself) the entities that can obtain certain context information.

○ **Obfuscation techniques.** Under this name we group the techniques, usually based on generalization or partial suppression, that limit the disclosure of private information contained in a request. Intuitively, they control the release of the right-hand part of the sensitive association (depicted in Figure 1).

○ **Identity anonymization techniques.** These are techniques that aim at avoiding the release of the left-hand part of the sensitive association, i.e., the identity of the issuer. The goal is to make the issuer indistinguishable among a sufficiently large number of individuals.

This classification may apply as well to defenses against LBS privacy threats, however our description of available approaches and solutions will be focused on those for more complex context-aware services. Sections 2, 3, 4, and 5 address each of the above types of defenses, respectively. Based on the weaknesses emerged from the analysis of the existing techniques, in Section 6 we advocate the use of a combined approach, present preliminary proposals, and illustrate the general characteristics that a comprehensive combined approach may have. Section 7 concludes the paper.

2   Network and Cryptographic Protocols

The development of context-aware services received impulse by technological progresses in the area of wireless communications, mobile devices, and sensors. The use of wireless channels, and more generally insecure channels, poses a first threat for users' privacy since it makes easier for an adversary to acquire service requests and responses by eavesdropping the communication or analyzing traffic on the network. In the literature, several models have been proposed for privacy preservation in context-aware systems. While some of them rely on a *centralized* architecture with a single trusted entity in charge of ensuring users' privacy, other models rely on a *decentralized* architecture in which mobile devices use direct communication channels with service providers. In both cases, two natural countermeasures for privacy attacks are: a) implement secure communication channels so that no third party can obtain requests/responses while they are in transit, and b) avoid the recognition of the client's network address, even by the service provider, which may be untrusted.

In order to protect point-to-point communications, in addition to standard wireless security, different cryptographic techniques can be applied. One possibility is clearly for applications to rely on SSL to encrypt communication; an alternative (or additional) possibility is to provide authentication, authorization and channel encryption through systems like Kerberos [12]. Kerberos is based on a centralized entity, *Key Distribution Center* (KDC), in charge of authenticating clients and servers in the network, and providing them with the keys needed for encrypting the communications. The centralized model that inspires Kerberos does not protect from attacks aimed at acquiring the control of the KDC entity. Specific solutions to communication protection also depend on the considered architecture and adversary's model, and are outside the scope of this paper.

One of the first solutions for achieving communications anonymity was the use of *Mix-nets* [13]. Mix-nets are networks composed of *mixes*, i.e. servers that relays encrypted data from sources to destinations. The scope of mixes is to avoid the association between incoming and outgoing messages. Each mix receives sets of encrypted communications coming from different senders, it decrypts messages, re-orders them and re-encrypts them before forwarding to the destinations.

Different approaches (e.g., [14,15]) aim at guaranteeing a certain degree of anonymity working at the IP level. The fundamental intuition of *Crowds* [14] is that the sender of a message is anonymous when she can be confused in a crowd of other potential senders. Hence, when a user wants to initiate a communication, she firstly sends the message to a member of a predetermined crowd that decides with a certain probability whether to forward the message to the destination, or to forward it to a randomly chosen member of the same crowd. Since the message is randomly exchanged among members of the crowd, even if an adversary intercepts the communication, the identity of the real senders remains anonymous.

The *Tarzan* system [15] adopted a solution based on a network overlay that clusters nodes in subnetworks called *domains* on the base of their IP addresses. The IP hiding is achieved by the substitution of the sender's IP address with the pseudonym corresponding to its domain. Moreover, when a node needs to send a packet, its communications are filtered by a special server called *mimic* that is in charge of *i)* substituting the IP address and other information that could reveal the sender identity with the adequate pseudonym, and *ii)* of setting a virtual path (*tunnel*) that guarantees the communication encryption.

Most solutions presented in the literature apply a combination of cryptographic techniques and routing protocols for IP hiding (e.g., TOR [16], which is extensively described in Chapter 4) to protect from eavesdropping over the communication channel. Onion Routing [17] implements both the features of IP hiding and message encryption. In order to preserve the sender's IP address, each message travels towards the receiver via a series of proxies, called *onion routers*, which choose the next component of the path setting an unpredictable route. Each router in the path removes one encryption level from the message before forwarding it to the next router.

A different application of a privacy-preserving routing protocol was presented by Al-Muhtadi et al. in [18]: the proposed solution has been designed for protecting a user's privacy while moving in smart environments. This solution is based on a hierarchy of trusted servers where the leaves, called *portals*, are aware of user's location, while internal nodes are aware of services provided by the environment. The user accesses the network through a portal and, according to her privacy preferences, she is assigned to an internal node, called *lighthouse*, that has the task of filtering and encrypting all the communications between the user and the service provider. The lighthouse does not know the user's position but is aware of the next hop in the server hierarchy composing the path to the user's portal. Similarly, the portal does not know which service the user is asking for, but it is aware of the path to the chosen lighthouse. The privacy preservation is achieved by decoupling position data from both the identity information and other context parameters. However, this approach requires the servers in the hierarchy to be trusted and it does not protect by privacy attacks performed by acquiring the control of one of the nodes in the structure.

The use of cryptographic techniques can also be extended to hide from the service provider the exact request parameters as well as the response. This approach has been proposed in the area of LBS where location information is often considered sensitive by users. In particular, solutions based on this approach aim at retrieving the nearest neighbor (NN) point of interest (*poi*) with respect to the user position at the time of the request.

A first solution was proposed by Atallah and Frikken in [5]: the authors propose a form of encrypted query processing combining the use of a data structure suited for managing spatial information with a cryptographic schema for secret sharing. On the server side, location data are handled through a *directed acyclic graph* (*DAG*), whose nodes correspond to Voronoi regions obtained by a tessellation of the space with respect to *pois* stored by the service provider. The query processing is performed according to the protocol proposed by Atallah and Du in [19], which allows a client to retrieve the correct Voronoi area without communicating its precise location. The drawback of this solution is that, in order to resolve a NN query, the user needs to send a number of queries that is proportional to the depth of the *DAG* instead of a single request. The consequent communication overhead impacts on the network traffic and on the response time, which are commonly considered important factors in mobile computing.

Recently, a cryptographic approach inspired by the Private Information Retrieval (PIR) field was proposed by Ghinita et al. in [6]. The service provider builds a Voronoi tessellation according to the stored *pois*, and superimposes on its top a regular grid of arbitrary granularity. In order to obtain the response to a NN query the privacy preservation mechanism relies on a PIR technique that is used for encrypting the user query, and for retrieving part of the location database without revealing spatial information. Some of the strong points of this solution are that location data are never disclosed; the user's identity is confused among identities of all users; and no trusted third party is needed to protect the users' privacy. However, since mobile devices are often characterized by limited computational capability, the query encryption and the answer processing performed at the client side have a strong impact on service response time, network and power consumption. In particular, when applied to context-aware services that perform the adaptation on a wide set of heterogeneous context data, this technique may result in unacceptable computation overhead both at the client and at the server side.

3 Access Control in Context-Aware Systems

Pervasive computing environments claim for techniques to control release of data and access to resources on the basis of the context of users, environment, and hardware/software entities. In general, the problem of access control [20] consists of deciding whether to authorize or not a requesting entity (*subject*) to perform a given *action* on a given resource (*object*). Access control mechanisms have been thoroughly studied in many fields, including operating systems, databases, and distributed systems. However, the characteristic features of pervasive environments introduce novel issues that must be taken into account for devising effective access control mechanisms. In particular, differently from centralized organizational domains, pervasive environments are characterized by the intrinsic decentralization of authorization decisions, since the object owners (users, services, infrastructures) are spread through the environment, and may adopt different policies regarding disclosure of private information. Hence, specific techniques to deal with the mobility and continuously changing context of the involved entities are needed to adapt authorizations to the current situation.

To this aim various techniques for context-aware access control have been recently proposed. Context-aware access control strategies fall in two main categories. The first category is the one of techniques aimed at granting or denying access to resources considering the context of the requesting user and of the resource (e.g., [21,22,23]). The second category is the one of techniques aimed at controlling the release of a user's context data on the basis of the context of the requesting entity and of the user herself. In this section we concentrate on techniques belonging to the latter category. Techniques belonging to the former category are presented in a different chapter of this book. We only mention that, since those techniques imply the release of users' context data to the access control mechanism, generally they also adopt strategies to enforce users' privacy policies.

Proposed context-aware access control mechanisms can be roughly classified in those that derive from *discretionary (DAC)* [24] and those that derive from *role-based (RBAC)* [25] access control. In DAC systems, the owner of each object is in charge of stating policies to determine the access privileges on the basis of the subject identity. These techniques are well suited to domains in which subjects do not belong to a structured organization (e.g., they are well suited to generic Internet services), since they are released from the burden of managing groups or roles of subjects. On the other hand, techniques based on RBAC (in which the access privileges depend on the subject role) are well suited to structured organization domains (like, e.g., hospitals, companies), since the definition of functional roles simplifies the management of access control policies. Other techniques related to access-control in context-aware systems include the use of access-rights graphs and hidden constraints (e.g., the technique proposed by Hengartner and Steenkiste in [26]) as well as *zero-knowledge proof theory* [27] (e.g., the technique proposed by Wang et al. in [28]). These are called *secret authorization* mechanisms, since they allow an entity to certify to a verifier the possession of private information (e.g., context data) revealing neither the authorization policies nor the secret data.

In the following we briefly describe the access control techniques for context-awareness derived from DAC and RBAC models, respectively.

Techniques derived from DAC. Even early approaches to discretionary access control allowed the expression of conditions to constrain permissions on the basis of the spatial and temporal characterization of the subject. More recently, access control techniques specifically addressed to the protection of location information (e.g., [29,30]) have been proposed. However, the richness and dynamics of contextual situations that may occur in pervasive and mobile computing environments claim for the definition of formal languages to express complex conditions on a multitude of context data, as well as sufficiently expressive languages to represent the context itself. To this aim, *Houdini* [31] provides a comprehensive formal framework to represent dynamic context data, integrate them from heterogeneous sources, and share context information on the basis of users' privacy policies. In particular, privacy policies can be expressed considering the context of the data owner (i.e., the user) and the context of the subject. As an example, a user of a service for locating friends could state a policy to disclose her current location to her friends only if her mood is *good* and her current activity is not *working*. Privacy policies in *Houdini* are expressed in a restricted logic programming language supporting rule chaining but no cycles. Rules preconditions express conditions on context data, while postconditions express permissions to access contextual information; reasoning with the resulting language has low computational complexity. Policy conflict resolution is based on explicit rule priorities.

Another relevant proposal, specifically addressed to the preservation of mobile customers privacy, was presented by Atluri and Shin in [32]. That work proposes an access control system aimed at controlling the release of private data based on time, location, and customer's preferences. For instance, a user could state a policy to disclose her location and profile information only during the weekend and if she is in a mall, and only in exchange for a discount coupon on items in her shopping list. The proposed solution is based on an intermediary infrastructure in charge of managing location and profiles of mobile users and to enforce their privacy policies. A specific index structure as well as algorithms are presented to efficiently enforce the proposed techniques.

Techniques derived from RBAC. Various proposals have been made to extend RBAC policies with contextual conditions (e.g., the one presented by Kumar et al. in [21]), and in particular with spatio-temporal constraints (e.g., the one presented by Atluri and Chun in [33]). More recently, this approach has been applied to the privacy protection of personal context data. A proposal in this sense is provided by the *UbiCOSM* middleware [34], which tackles the comprehensive issue with mechanisms to secure the access not only to services provided by ubiquitous infrastructures, but also to users' context data, based on contextual conditions and roles. The context model of UbiCOSM distinguishes between the *physical* dimension, which describes the spatial characterization of the user, and the *logical* dimension, which describes other data such as the user's current activity and device capabilities. For instance, the context *TouristAtMuseum* is composed of the physical context *AtMuseum* (characterized by the presence of the user within the physical boundaries of a museum) and by the logical context *Tourist* (which defines the user's role as the one of a tourist). Users can declare a policy to control the release of personal context data as the association between a permission and a context in which the permission applies. Simple context descriptions can be composed in more complex ones by means of logical operators, and may involve the situation of multiple entities. For instance, in order to find other tourists that share her same interests, a user could state a policy to disclose her cultural preferences to a person only if their current context is *TouristAtMuseum* and they are both co-located with a person that is a friend of both of them.

Another worth-mentioning system is *CoPS* [35], which provides fine-grained mechanisms to control the release of personal context data, as well as techniques to identify misuse of the provided information. In particular, policies in CoPS are organized in a hierarchical manner, on the basis of the priority level of the policy (i.e., organization-level, user-level, default). Permissions depend on the context and the role of the subject. CoPS supports both administrator and user-defined roles. While the former reflect the hierarchical structure of the organization, the latter can be used to categorize entities in groups, in order to simplify the policy management by users. The system adopts a conflict resolution mechanism based on priorities and on the specificity of access control rules. Moreover, a trigger mechanism can be set up to control the release of particular context data against the frequency of the updates; this technique can be used, for instance, to notify the user in the case someone tries to track her movements by continuously polling her location.

Open issues and remarks. As emerged from the above analysis of the state-of-the-art, the main strong point of techniques derived from DAC consists in the efficiency of the reasoning procedures they employ to evaluate at run-time the access privileges of the requesting entity. This characteristic makes them very well suited to application domains characterized by strict real-time requirements. On the other hand, the roles abstraction adopted by techniques derived from RBAC can be profitably exploited not only in structured organizational domains but also in open environments (like ambient intelligence systems), since heterogeneous entities can be automatically mapped to predefined roles on the basis of the contextual situation to determine their access privileges.

Nevertheless, some open issues about context-aware access control systems are worth to be considered. In particular, like in generic access control systems, a formal model to represent policies and automatically recognize inconsistencies (especially in systems supporting the definition of negative authorizations) is needed; however, only part of the techniques proposed for context-aware computing face this issue. This problem is further complicated by the fact that the privacy policy of a subject may conflict with the privacy policy of an object owner. Proposed solutions for this issue include the use of techniques for secret authorization, like proposed by Hengartner and Steenkiste in [26]. Moreover, an evident weakness of these systems consists in their rigidity: if strictly applied, an access control policy either grants or denies access to a given object. This weakness is alleviated by the use of obfuscation techniques (reported in Section 4) to disclose the required data at different levels of accuracy on the basis of the current situation.

A further critical issue for context-aware access control systems consists in devising techniques to support end users in defining privacy policies. Indeed, manual policy definition by users is an error-prone and tedious task. For this reason, straightforward techniques to support users' policy definition consists in making use of user friendly interfaces and default policies, like in Houdini and in CoPS, respectively. However, a more sophisticated strategy to address this problem consists in the adoption of statistical techniques to automatically learn privacy policies on the basis of the past decisions of the user. To this aim, Zhang et al. propose in [36] the application of rough set theory to extract access control policies based on the observation of the user's interaction with context-aware applications during a training period.

As a final remark, we point out that context-aware access control systems do not protect privacy in the case the access to a service is considered private information by itself (e.g., because it reveals particular interests or habits about the user). To address this issue, techniques aimed at enforcing anonymity exist and are reviewed in Section 5.

4   Obfuscation of Context Data

Access control systems either deny or allow access to a given context data depending on the current situation. For instance, consider the user of a service that redirects incoming calls and messages on the basis of the current activity. Suppose that the service is not completely trusted by the user; hence, since she considers her current activity (e.g., *MeetingCustomers*) a sensitive information, whether to allow or deny the access to her precise current activity may be unsatisfactory. Indeed, denying access to those data would determine the impossibility to take advantage of that service, while allowing access could result in a privacy violation. In this case, a more flexible solution is to *obfuscate* [37] the private data before communicating them to the service provider in order to decrease the sensitivity level of the data. For instance, the precise current activity *Meeting-Customers* could be obfuscated to the more generic activity *BusinessMeeting*. This solution is based on the intuition that each private information is associated to a given sensitivity level, which depends on the precision of the information itself; generally, the lesser the information is precise, the lesser it is sensitive. Obfuscation techniques have been applied to the protection of microdata released from databases (e.g., the technique proposed by Xiao and Tao in [38]).

Several techniques based on obfuscation have also been proposed to preserve the privacy of users of context-aware services. These techniques are generally coupled with an access control mechanism to tailor the obfuscation level to be enforced according to the trustiness of the subject and to the contextual situation. However, in this section we concentrate on works that specifically address context data obfuscation. The main research issue in this field is to devise techniques to provide adequate privacy preservation while retaining the usefulness of the data to context-awareness purposes. We point out that, differently from techniques based on anonymity (reviewed in Section 5), techniques considered in this section do not protect against the disclosure of the user's identity.

Various obfuscation-based techniques to control the release of location information have been recently proposed (e.g., [7,8,39]), based on generalization or perturbation of the precise user's position. One of the first attempts to support privacy in generic context aware systems through obfuscation mechanisms is *semantic eWallet* [40], an architecture to support context-awareness by means of techniques to retrieve users' context data while enforcing their privacy preferences. Users of the semantic eWallet may express their preferences about the accuracy level of their context data based on the requester's identity and on the context of the request. That system supports both *abstraction* and *falsification* of context information. By abstraction, the user can decide to generalize the provided data, or to omit some details about it. For instance, a user involved in a *BusinessMeeting* could decide to disclose her precise activity to a colleague only during working hours and if they both are located within a company building; activity should be generalized to *Meeting* in the other cases. On the other hand, by falsification the user can decide to deliberately provide false information in order to mask her precise current context in certain situations. For instance, a CEO could reveal to her secretary that she is currently *AtTheDentist*, while telling the other employees that she is involved in a *BusinessMeeting*. In the semantic eWallet, context data are represented by means of ontologies. Obfuscation preferences are encoded as rules whose preconditions include a precise context data and conditions for obfuscation, and postconditions express the obfuscated context data to be disclosed if the preconditions hold.

While in the semantic eWallet the mapping between precise and obfuscated information must be explicitly stated case-by-case, a more scalable approach to the definition of obfuscation preferences is proposed by Wishart et al. in [41]. That work copes with the multi-party ownership of context information in pervasive environments by proposing a framework to retrieve context information and distributing it on the basis of the obfuscation preferences stated by the data owner. It is worth to note that in the proposed framework the owner of the data is not necessarily the actual proprietary of the context source; instead, the data owner is the person whom the data refer to. For instance, the owner of data provided by a server-side positioning system is the user, not the manager of the positioning infrastructure; hence, the definition of obfuscation preferences about personal location is left to the user. Obfuscation preferences are expressed by conditions on the current context, by specific context data, and by a maximum detail level at which those data can be disclosed in that context. The level of detail of context data refers to the specificity of that data according to a predefined *obfuscation ontology*. Context data in an obfuscation ontology are organized as nodes into a hierarchy, such that parent nodes represent more general concepts with respect to their children; e.g., the activity *MeetingCustomers* has parent activity *BusinessMeeting*, which in turn has parent activity *Working*. For instance, an obfuscation preference could state to disclose the user's current activity with a *level 2* specificity in the case the requester is *Bob* and the request is made during *working hours*. In the case those conditions hold, the released data are calculated by generalizing the exact current activity up to the second level of the *Activity* obfuscation ontology (i.e., up to the level of the grandchildren of the root node), or to a lower level if the available information is less specific than that stated by the preference. Since manually organizing context data in an obfuscation ontology could be unpractical, a technique to automatically discover reasoning modules able to derive the data at the required specificity level is also presented.

Based on the consideration that the quality of context information (QoC) is a strong indicator of privacy sensitiveness, Sheikh et al. propose the use of QoC to enforce users' privacy preferences [42]. In that work, the actual quality of the disclosed context data is negotiated between service providers and users. When a service provider needs data regarding a user's context, it specifies the QoC that it needs for those data in order to provide the service. On the other hand, the user specifies the maximum QoC she is willing to disclose for those data in order to take advantage of the service. Service requirements and user's privacy preferences are communicated to a middleware that is in charge of verifying if they are incompatible (i.e., if the service requires data to a quality the user is not willing to provide). If this is not the case, obfuscation mechanisms are applied on those data in order to reach the quality level required by the service provider. QoC is specified on the basis of five indicators, i.e., precision, freshness, spatial and temporal resolution, and probability of correctness. Each context data are associated with five numerical values that express the quality of the data with respect to each of the five indicators. Given a particular context situation, a user can specify her privacy preferences for context data by defining the maximum quality level for each of the five indicators that she is willing to disclose in that situation. For instance, the user of a remote health monitoring service could state to disclose vague context information to the caregivers when in a non-emergency context, while providing accurate data in the case of emergency.

One inherent weakness of obfuscation techniques for privacy in context-awareness is evident: if the service provider requires context data to a quality that the user is not willing to disclose, access to that service is not possible. In order to overcome this issue, anonymization techniques (presented in Section 5) have been proposed, which protect from the disclosure of the user's identity, while possibly providing accurate context information.

5   Identity Anonymization Techniques

While obfuscation techniques aim at protecting the right-hand side of the sensitive association (SA) (see Figure 1), the goal of techniques for identity anonymization is to protect the left-hand side of the SA in order to avoid that an adversary re-identifies the issuer of a request.

In the area of database systems, the notion of k-anonymity has been introduced by Samarati in [43] to formally define when, upon release of a certain database view containing records about individuals, for any specific sensitive set of data in the view, the corresponding individual can be considered indistinguishable among at least k individuals. In order to enforce anonymity it is necessary to determine which attributes in a table play the role of *quasi-identifiers* (QI), i.e., data that joined with external knowledge may help the adversary to restrict the set of candidate individuals. Techniques for database anonymization adopt generalization of QI values and/or suppression of records in order to guarantee that be partitioned in groups of at least k records having the same value for QI attributes (called QI-*groups*). Since each individual is assumed to be the respondent of a single record, this implies that there are at least k candidate respondents for each released record.

The idea of k-anonymity has also been applied to define a privacy metric in location based services, as a specific kind of context-aware services (for instance, as done by Gruteser and Grunwald in [9]). In this case, the information being released is considered the information in the service request. In particular, the information about the user's location may be used by an adversary to re-identify the issuer of the request if the adversary has access to external information about users' location. Attacks and defense techniques in this context have been investigated in several papers, among which [9,10]. Moreover, a formal framework for the categorization of defense techniques with respect to the adversary's knowledge assumptions has been proposed by Bettini et al. in [3]. According to that categorization, when the adversary performs his attack using information contained in a single request the attack is said to be *single-issuer*; otherwise, when the adversary may compare information included in requests by multiple users, the attack is said to be *multiple-issuers*. Moreover, cases in which the adversary can acquire information only during a single time granule are called static (or *snapshot*), while contexts in which the adversary may observe multiple requests issued by the same users in different time granules are called dynamic (or *historical*). A possible technique to enforce anonymity in LBS is to generalize precise location data in a request to an area including a set (called *anonymity set* [44]) of other potential issuers. An important difference between the anonymity set in service requests and the *QI-group* in databases is that while the *QI-group* includes only identities actually associated to a record in the table, the anonymity set includes also users that did not issue any request but that are potential issuers with respect to the adversary's external knowledge.

With respect to identity anonymization in generic context-aware systems, it is evident that many other kinds of context data besides location may be considered *QI*. Hence, a large amount of context data must be generalized in order to enforce anonymity. As a consequence, the granularity of generalized context data released to the service provider could be too coarse to provide the service at an acceptable quality level. In order to limit the information loss due to the generalization of context data, four different personalized anonymization models are proposed by Shin et al. in [45]. These models allow a user to constrain the maximum level of location and profile generalization still guaranteeing the desired level of anonymity. For instance, a user could decide to constrain the maximum level of location generalization to an area of $1\ km^2$, while imposing no constraints on the level of generalization of her profile.

As outlined in the introduction, sensing technologies deployed in pervasive environments can be exploited by adversaries to constantly monitor the users' behavior, thus exposing the user to novel kinds of privacy attacks, like the one presented by Riboni et al. in [4]. In that work it is shown that even enforcing k-anonymity, in particular cases the attacker may recognize the actual issuer of a service request by monitoring the behavior of the potential issuers with respect to service responses. For example, consider a pervasive system of a gym, suggesting exercises on the basis of gender, age, and physiological data retrieved from body-worn sensors. Even if users are anonymous in a set of k potential issuers, the attacker can easily recognize the issuer of a particular request if she starts to use in a reasonable lapse of time a machine the system suggested to her, which was not suggested to any other potential issuer. The proposed solution relies on an intermediary entity that filters all the communications between users and service providers, calculates the privacy threats corresponding to possible alternatives suggested by the service (e.g., the next exercise to perform), and automatically filters unsafe alternatives.

In the scenario considered by Hore et al. in [46], context data are used for automatically detecting complex events in pervasive environments. Since complex events are defined as compositions of simple events, and each simple event has an associated set of possible participants, the intersection of those sets may lead to the identification of the actual actors. Hence, in order to preserve users' privacy, the authors propose a technique to guarantee that each complex event has at least k possible participants.

A further issue to be considered is the defense against the well-known problem of *homogeneity* [47] identified in the field of databases. Homogeneity attacks can be performed if all the records belonging to a qi-group have the same value of sensitive information. In this case it is clear that the adversary may easily violate the users' privacy despite anonymity is formally enforced. The same problem may arise as well in context-aware services in the case an adversary recognizes that all the users in an anonymity set actually issued a request with the same value of private information. To our knowledge, a first effort to defend against such attacks in context-aware systems has been presented by Riboni et al. in [48]. That proposal aims at protecting from multiple-issuers historical attacks by applying a bounded generalization of both context data and service parameters.

6   Towards a Comprehensive Framework for Privacy Protection in Context-Aware Systems

Based on the weaknesses emerged from the analysis of the proposed techniques, in this section we advocate the use of a combined approach to address the comprehensive issue of privacy in context awareness; we present existing proposals, and we illustrate the logical design of a framework intended to solve most of the identified problems.

A reference scenario. In order to illustrate the weaknesses of existing techniques we present the following reference scenario.

Scenario: *Consider the pervasive system of a sports center in which users wear a smart watch that collects context data from body-worn sensors to continuously monitor data such as user's position, the used equipments, and physiological parameters. These data are communicated from users to a* virtual trainer *service of the sports center included in a request to obtain suggestions for the next exercise/activity. Since physiological data are particularly sensitive (because they can reveal important details about a person's health status), their association to the actual owners needs to be protected from untrusted entities such as the system of the sports center. The system is also able to collect a subset of the users' context data; in particular, it can continuously monitor the users' positions. Since it knows users' identities, their position, and the map of the center, the system is anytime aware of who is using a given equipment or performing a given activity.*

On the need for a combined approach. The analysis of the state-of-the-art reported in the previous sections has shown that each of the proposed approaches, even if effective in a particular scenario and under particular assumptions, fails in providing a solution to the general problem. In particular, considering our reference scenario, we note that:

○ cryptographic techniques for private information retrieval presented up to the time of writing are unfeasible to support complex context-aware services like the virtual trainer service, due to problems of bandwidth and computational resources consumption;

○ protecting communication privacy between the context source and the context data consumer (e.g., the virtual trainer service) is useless in the case the context data consumer is untrusted. In order to protect users' privacy, they must be coupled with techniques for anonymity/obfuscation;

○ access control techniques (possibly coupled with obfuscation) do not prevent a malicious subject from adopting reasoning techniques in order to derive new sensitive information based on data it is authorized to access. For instance, in the considered scenario physiological parameters of a user could be statistically analyzed by the owner of the virtual trainer service to derive sensitive information about the user's health status. Hence, in this case access control techniques should be coupled with techniques for enforcing identity anonymity;

○ techniques for identity anonymity rely on the exact knowledge about the external information available to an adversary. However, especially in pervasive and mobile computing scenarios, such knowledge is very hard to obtain, and adopting worst-case assumptions about the external information leads to a significant degradation of the quality of released context data. Moreover, as shown by Riboni et al. in [4], in pervasive environments these techniques are prone to attacks based on the observation of the behavior of service users.

These observations claim for the combination of different approaches in order to protect against the different kind of attacks that can be posed to the privacy of users taking advantage of context-aware services.

Existing techniques. Proposals to combine different approaches in a common framework have been recently presented.

An architecture for privacy-conscious context aggregation and reasoning is proposed by Pareschi et al. in [49]. The proposed solution adopts client-side reasoning modules to abstract raw context data into significant descriptions of the user's situation (e.g., current activity and stereotype) that can be useful for adaptation. Release of private context information is controlled by context-aware access control policies, and the access to context information by service providers is mediated by a trusted intermediary infrastructure in charge of enforcing anonymity. Moreover, cryptographic techniques are used to protect communications inside the user trusted domain.

Papadopoulou et al. present in [50] a practical solution to enforce anonymity. In that work, no assumptions about the external knowledge available to an adversary are made; hence, the proposed technique does not formally guarantee a given anonymity level. For this reason, the anonymization technique is coupled with access control and obfuscation mechanisms in order to protect privacy in the case an adversary is able to discover the user's identity. That technique is applied using the *virtual identity* metaphor. A virtual identity is essentially the subset of context data that a user is willing to share with a third party in a given situation; in addition, since anonymity is not formally guaranteed, part of the shared context data can be obfuscated on the basis of privacy policies in order to hide some sensible details. For instance, a person could decide to share her preferences regarding shopping items and leisure activities, as well as her obfuscated location, when she is on vacation (using a *tourist* virtual identity), while hiding those information when she is traveling for work (using a *worker* virtual identity). With respect to the problem introduced by multiple requests issued by the same user, specific techniques are presented to avoid that different virtual identities can be linked to the same (anonymous) user by an adversary.

While the above mentioned works try to protect the privacy of users accessing a remote service, the *AnonySense* system [51] is aimed at supporting privacy in opportunistic sensing applications, i.e., applications that leverage opportunistic networks formed by mobile devices to acquire aggregated context data in a particular region. To reach this goal, the geographic area is logically partitioned into tiles large enough to probabilistically gain k-anonymity; i.e., regions visited with high probability by more than k persons during a given time granule. Measurements of context data are reported by mobile nodes specifying the tile they refer to and the time interval during which they were acquired. Moreover, in order to provide a second layer of privacy protection, obfuscation is applied on the sensed data by fusing the values provided by at least l nodes ($l \leq k$) before communicating the aggregated data to the application. Cryptographic techniques are used to enforce anonymous authentication by users of the system.

Towards a comprehensive framework. We now illustrate how existing techniques can be extended and combined in a logical multilayer framework, which is graphically depicted in Figure 2. This framework is partially derived from the preliminary architecture described by Pareschi et al. in [49]. However, the model presented here is intended to provide a more comprehensive privacy solution, addressing problems regarding sensor and profile data aggregation and reasoning (including obfuscation), context-aware access control and secret authorization, anonymous authentication, identity anonymity, and anonymous/encrypted communication. Clearly, the actual techniques to be applied for protecting privacy depend on the current context (users' situation, available services, network and environmental conditions). However, we believe that this framework is flexible enough to provide effective privacy protection in most pervasive and mobile computing scenarios. The framework is composed of the following layers:

○ ***Sensors* layer:** This layer includes body-worn and environmental sensors that communicate context data to the upper layers through encrypted channels using energy-efficient cryptographic protocols (e.g., those based on elliptic curves [52] like in Sun SPOT sensors [53]). We assume that this layer

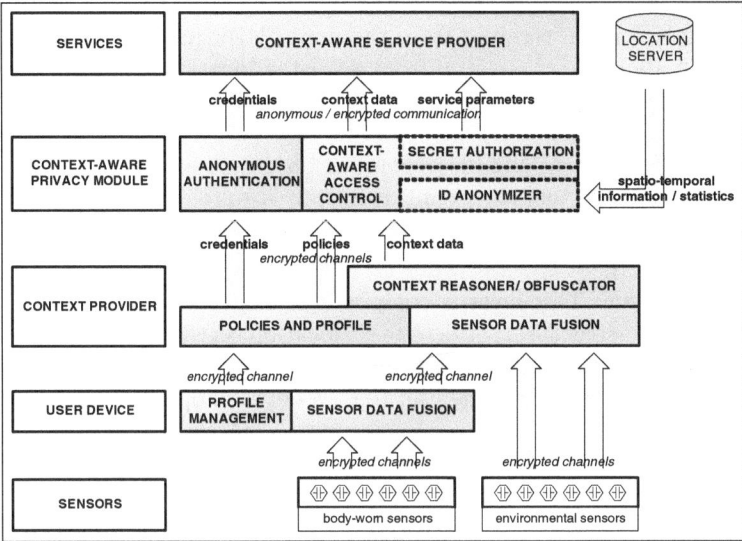

Fig. 2. The envisioned framework

is within the trusted domain of the user (i.e., sensors do not deliberately provide false information).

○ **User device layer:** This layer is in charge of managing the user's profile information (i.e., context data that are almost static, like personal information, interests and preferences) and privacy policies. Upon update of this information by the user, the new information is communicated to the upper layer. Moreover, this layer is in charge of fusing context data provided by body-worn sensors and to communicate them in an aggregated form to the upper layer on a *per-request* basis. This layer is deployed on the user's device, which is assumed to be trusted (traditional security issues are not addressed here); communications with the upper layer are performed through encrypted channels.

○ **Context provider layer:** This layer is in charge of fusing sensor data provided by the lower layers, including those provided by sensors that are not directly under the communication range of the user device. Moreover, according to the user's policies, it performs context reasoning and obfuscation for privacy and adaptation purposes, as described by Pareschi et al. in [49]. It communicates user's credentials, privacy policies, and context data to the upper layer on a *per-request* basis through encrypted channels. This layer belongs to the user's trusted domain; depending on the device capabilities, it can be deployed on the user's device itself, or on another trusted machine.

○ **Context-aware privacy module layer:** This layer is in charge of anonymously authenticating the user on the upper layer, and to enforce her context-aware access control policies, possibly after a phase of secret negotiation with the third party. Moreover, depending on the user's policies, it can possibly anonymize the user's identity on the basis of (either precise or statistical) trusted information received from the upper layer (e.g., spatio-temporal information about users received from a trusted location server). Protocols for anonymous/encrypted communication are adopted to provide credentials, context data and service parameters to the upper layer. This layer belongs to the user's trusted domain. Depending on device capabilities and on characteristics of the actual algorithms it adopts (e.g., to enforce anonymity), this layer can be implemented on the user's device, on another trusted machine, or on the infrastructure of a trusted entity (e.g., the network operator).

○ **Services layer:** This layer is composed of context-aware service providers and other infrastructural services (e.g., location servers). Typically, this layer is assumed not to belong to the user's trusted domain, even if particular services can be trusted by the user (e.g., a network operator location server).

7 Conclusions

Through a classification into four main categories of techniques, we have described the state of the art of privacy preservation for georeferenced context-aware services. While previous work has also proposed the combination of techniques from two or more categories, we claim that a deeper integration is needed and we propose an architecture for a comprehensive framework towards this goal. Clearly, there is still a long way to go in order to refine the architecture, work out the details of its components, implement and integrate the actual techniques, and test the framework on real applications. Moreover, there are still several other aspects, not considered in our paper, that deserve investigation. For example, since there are well-known techniques for context reasoning, they may have to be taken into account, since released context data may determine the disclosure of other context data, possibly leading to privacy leaks that were previously unidentified. Furthermore, computationally expensive techniques (e.g., those making use of ontological reasoning or complex cryptographic algorithms) pose serious scalability issues that may limit their applicability in real-world scenarios. Finally, since the access to context data of real users is generally unavailable for privacy reasons, sophisticated simulation environments are needed to evaluate the actual effectiveness of privacy preservation mechanisms in realistic situations.

Acknowledgments

This work was partially supported by National Science Foundation (NSF) under grant N. CNS-0716567, and by Italian MIUR under grants InterLink II04C0EC1D and PRIN-2007F9437X.

References

1. Palen, L., Dourish, P.: Unpacking "privacy" for a networked world. In: Proceedings of the 2003 Conference on Human Factors in Computing Systems (CHI 2003), pp. 129–136. ACM Press, New York (2003)

2. Lederer, S., Hong, J.I., Dey, A.K., Landay, J.A.: Personal privacy through under-standing and action: five pitfalls for designers. Personal and Ubiquitous Computing 8(6), 440–454 (2004)
3. Bettini, C., Mascetti, S., Wang, X.S.: Privacy Protection through Anonymity in Location-based Services. In: Handbook of Database Security: Applications and Trends, pp. 509–530 (2008)
4. Riboni, D., Pareschi, L., Bettini, C.: Shadow attacks on users' anonymity in pervasive computing environments. Pervasive and Mobile Computing 4(6), 819–835 (2008)
5. Atallah, M.J., Frikken, K.B.: Privacy-Preserving Location-Dependent Query Processing. In: ICPS 2004: Proceedings of the The IEEE/ACS International Conference on Pervasive Services, pp. 9–17. IEEE Computer Society Press, Los Alamitos (2004)
6. Ghinita, G., Kalnis, P., Khoshgozaran, A., Shahabi, C., Tan, K.L.: Private queries in location based services: anonymizers are not necessary. In: Proceedings of the ACM SIGMOD International Conference on Management of Data (SIGMOD 2008), pp. 121–132. ACM Press, New York (2008)
7. Ardagna, C.A., Cremonini, M., Damiani, E., De Capitani di Vimercati, S., Samarati, P.: Location Privacy Protection Through Obfuscation-Based Techniques. In: Barker, S., Ahn, G.-J. (eds.) Data and Applications Security 2007. LNCS, vol. 4602, pp. 47–60. Springer, Heidelberg (2007)
8. Yiu, M.L., Jensen, C.S., Huang, X., Lu, H.: SpaceTwist: Managing the Trade-Offs Among Location Privacy, Query Performance, and Query Accuracy in Mobile Services. In: Proceedings of the 24th International Conference on Data Engineering (ICDE 2008), pp. 366–375. IEEE Computer Society Press, Los Alamitos (2008)
9. Gruteser, M., Grunwald, D.: Anonymous Usage of Location-Based Services Through Spatial and Temporal Cloaking. In: Proc. of the 1st International Conference on Mobile Systems, Applications and Services (MobiSys), pp. 31–42. USENIX Association (2003)
10. Gedik, B., Liu, L.: Protecting Location Privacy with Personalized k-Anonymity: Architecture and Algorithms. IEEE Transactions on Mobile Computing 7(1), 1–18 (2008)
11. Aggarwal, C.C.: On k-Anonymity and the Curse of Dimensionality. In: Proceedings of the 31st International Conference on Very Large Data Bases (VLDB), pp. 901–909. ACM Press, New York (2005)
12. Neuman, B., Ts'o, T.: Kerberos: an authentication service for computer networks. IEEE Communications Magazine 32(9), 33–38 (1994)
13. Chaum, D.: Untraceable electronic mail, return addresses, and digital pseudonyms. Commun. ACM 24(2), 84–90 (1981)
14. Reiter, M.K., Rubin, A.D.: Anonymous web transactions with crowds. Commun. ACM 42(2), 32–48 (1999)
15. Freedman, M.J., Morris, R.: Tarzan: a peer-to-peer anonymizing network layer. In: CCS 2002: Proceedings of the 9th ACM conference on Computer and communications security, pp. 193–206. ACM Press, New York (2002)
16. Dingledine, R., Mathewson, N., Syverson, P.: Tor: the second-generation onion router. In: SSYM 2004: Proceedings of the 13th conference on USENIX Security Symposium, p. 21. USENIX Association (2004)
17. Goldschlag, D., Reed, M., Syverson, P.: Onion routing. Commun. ACM 42(2), 39–41 (1999)

18. Al-Muhtadi, J., Campbell, R., Kapadia, A., Mickunas, M.D., Yi, S.: Routing Through the Mist: Privacy Preserving Communication in Ubiquitous Computing Environments. In: Proceedings of the 22 nd International Conference on Distributed Computing Systems (ICDCS 2002), p. 74. IEEE Computer Society Press, Los Alamitos (2002)
19. Atallah, M.J., Du, W.: Secure multi-party computational geometry. In: Dehne, F., Sack, J.-R., Tamassia, R. (eds.) WADS 2001. LNCS, vol. 2125, pp. 165–179. Springer, Heidelberg (2001)
20. Samarati, P., De Capitani di Vimercati, S.: Access Control: Policies, Models, and Mechanisms. In: Focardi, R., Gorrieri, R. (eds.) FOSAD 2000, vol. 2171, pp. 137–196. Springer, Heidelberg (2001)
21. Kumar, A., Karnik, N.M., Chafle, G.: Context sensitivity in role-based access control. Operating Systems Review 36(3), 53–66 (2002)
22. Covington, M.J., Fogla, P., Zhan, Z., Ahamad, M.: A Context-Aware Security Architecture for Emerging Applications. In: Proceedings of the 18th Annual Computer Security Applications Conference (ACSAC 2002), pp. 249–260. IEEE Computer Society Press, Los Alamitos (2002)
23. Toninelli, A., Montanari, R., Kagal, L., Lassila, O.: Proteus: A Semantic Context-Aware Adaptive Policy Model. In: Proceedings of the 8th IEEE International Workshop on Policies for Distributed Systems and Networks(POLICY 2007), pp. 129–140. IEEE Computer Society, Los Alamitos (2007)
24. Sandhu, R., Samarati, P.: Access Control: Principles and Practice. IEEE Communications 32(9), 40–48 (1994)
25. Sandhu, R.S., Coyne, E.J., Feinstein, H.L., Youman, C.E.: Role-Based Access Control Models. IEEE Computer 29(2), 38–47 (1996)
26. Hengartner, U., Steenkiste, P.: Avoiding Privacy Violations Caused by Context-Sensitive Services. Pervasive and Mobile Computing 2(3), 427–452 (2006)
27. Goldwasser, S., Micali, S., Rackoff, C.: The knowledge complexity of interactive proof systems. SIAM Journal of Computing 18(1), 186–208 (1989)
28. Wang, C.D., Feng, L.C., Wang, Q.: Zero-Knowledge-Based User Authentication Technique in Context-aware System. In: International Conference on Multimedia and Ubiquitous Engineering, 2007. MUE 2007, pp. 874–879 (2007)
29. Hengartner, U., Steenkiste, P.: Access control to people location information. ACM Trans. Inf. Syst. Secur. 8(4), 424–456 (2005)
30. Schulzrinne, H., Tschofenig, H., Morris, J., Cuellar, J., Polk, J.: Geolocation policy: A document format for expressing privacy preferences for location information (January 2009),
    http://www.ietf.org/internet-drafts/draft-ietf-geopriv-policy-18.txt
31. Hull, R., Kumar, B., Lieuwen, D., Patel-Schneider, P., Sahuguet, A., Varadarajan, S., Vyas, A.: Enabling Context-Aware and Privacy-Conscious User Data Sharing. In: Proceedings of the 2004 IEEE International Conference on Mobile Data Management (MDM 2004), pp. 187–198. IEEE Computer Society, Los Alamitos (2004)
32. Atluri, V., Shin, H.: Efficient Security Policy Enforcement in a Location Based Service Environment. In: Barker, S., Ahn, G.-J. (eds.) Data and Applications Security 2007. LNCS, vol. 4602, pp. 61–76. Springer, Heidelberg (2007)
33. Atluri, V., Chun, S.A.: A geotemporal role-based authorisation system. International Journal of Information and Computer Security 1(1–2), 143–168 (2007)
34. Corradi, A., Montanari, R., Tibaldi, D.: Context-Based Access Control Management in Ubiquitous Environments. In: Proceedings of the 3rd IEEE International Symposium on Network Computing and Applications (NCA 2004), pp. 253–260. IEEE Computer Society Press, Los Alamitos (2004)

35. Sacramento, V., Endler, M., Nascimento, F.N.: A Privacy Service for Context-aware Mobile Computing. In: Proceedings of the First International Conference on Security and Privacy for Emerging Areas in Communications Networks (SECURECOMM 2005), pp. 182–193. IEEE Computer Society, Los Alamitos (2005)
36. Zhang, Q., Qi, Y., Zhao, J., Hou, D., Zhao, T., Liu, L.: A Study on Context-aware Privacy Protection for Personal Information. In: Proceedings of the 16th IEEE International Conference on Computer Communications and Networks (ICCCN 2007), pp. 1351–1358. IEEE Computer Society, Los Alamitos (2007)
37. Bakken, D.E., Parameswaran, R., Blough, D.M., Franz, A.A., Palmer, T.J.: Data Obfuscation: Anonymity and Desensitization of Usable Data Sets. IEEE Security & Privacy 2(6), 34–41 (2004)
38. Xiao, X., Tao, Y.: Personalized privacy preservation. In: SIGMOD 2006: Proceedings of the 2006 ACM SIGMOD international conference on Management of data, pp. 229–240. ACM Press, New York (2006)
39. Duckham, M., Kulik, L.: A Formal Model of Obfuscation and Negotiation for Location Privacy. In: Gellersen, H.-W., Want, R., Schmidt, A. (eds.) Pervasive 2005. LNCS, vol. 3468, pp. 152–170. Springer, Heidelberg (2005)
40. Gandon, F.L., Sadeh, N.M.: Semantic web technologies to reconcile privacy and context awareness. J. Web Sem. 1(3), 241–260 (2004)
41. Wishart, R., Henricksen, K., Indulska, J.: Context Privacy and Obfuscation Supported by Dynamic Context Source Discovery and Processing in a Context Management System. In: Indulska, J., Ma, J., Yang, L.T., Ungerer, T., Cao, J. (eds.) UIC 2007. LNCS, vol. 4611, pp. 929–940. Springer, Heidelberg (2007)
42. Sheikh, K., Wegdam, M., van Sinderen, M.: Quality-of-Context and its use for Protecting Privacy in Context Aware Systems. Journal of Software 3(3), 83–93 (2008)
43. Samarati, P.: Protecting Respondents' Identities in Microdata Release. IEEE Trans. on Knowledge and Data Engineering 13(6), 1010–1027 (2001)
44. Pfitzmann, A., Köhntopp, M.: Anonymity, unobservability, and pseudonymity - a proposal for terminology. In: Federrath, H. (ed.) Designing Privacy Enhancing Technologies. LNCS, vol. 2009, pp. 1–9. Springer, Heidelberg (2001)
45. Shin, H., Atluri, V., Vaidya, J.: A Profile Anonymization Model for Privacy in a Personalized Location Based Service Environment. In: Proceedings of the 9th International Conference on Mobile Data Management (MDM 2008), pp. 73–80 (2008)
46. Hore, B., Wickramasuriya, J., Mehrotra, S., Venkatasubramanian, N., Massaguer, D.: Privacy-Preserving Event Detection in Pervasive Spaces. In: Proceedings of the 7th IEEE International Conference on Pervasive Computing and Communications (PerCom 2009) (2009)
47. Machanavajjhala, A., Gehrke, J., Kifer, D., Venkitasubramaniam, M.: l-Diversity: Privacy Beyond k-Anonymity. In: Proceedings of ICDE 2006. IEEE Computer Society, Los Alamitos (2006)
48. Riboni, D., Pareschi, L., Bettini, C., Jajodia, S.: Preserving Anonymity of Recurrent Location-based Queries. In: Proceedings of the 16th International Symposium on Temporal Representation and Reasoning (TIME 2009) (to appear, 2009)
49. Pareschi, L., Riboni, D., Agostini, A., Bettini, C.: Composition and Generalization of Context Data for Privacy Preservation. In: Sixth Annual IEEE International Conference on Pervasive Computing and Communications (PerCom 2008), Proceedings of the Workshops, pp. 429–433. IEEE Computer Society, Los Alamitos (2008)

50. Papadopoulou, E., McBurney, S., Taylor, N., Williams, M.H., Dolinar, K., Neubauer, M.: Using User Preferences to Enhance Privacy in Pervasive Systems. In: Proceedings of the Third International Conference on Systems (ICONS 2008), pp. 271–276. IEEE Computer Society, Los Alamitos (2008)

51. Kapadia, A., Triandopoulos, N., Cornelius, C., Peebles, D., Kotz, D.: AnonySense: Opportunistic and Privacy-Preserving Context Collection. In: Indulska, J., Patterson, D.J., Rodden, T., Ott, M. (eds.) Pervasive 2008. LNCS, vol. 5013, pp. 280–297. Springer, Heidelberg (2008)

52. Miller, V.S.: Use of Elliptic Curves in Cryptography. In: Williams, H.C. (ed.) CRYPTO 1985. LNCS, vol. 218, pp. 417–426. Springer, Heidelberg (1986)

53. Simon, D., Cifuentes, C., Cleal, D., Daniels, J., White, D.: Java^{TM} on the bare metal of wireless sensor devices: the squawk Java virtual machine. In: Proceedings of the 2nd International Conference on Virtual Execution Environments (VEE 2006), pp. 78–88. ACM Press, New York (2006)

Safety and Privacy in Vehicular Communications

Josep Domingo-Ferrer and Qianhong Wu

Universitat Rovira i Virgili,
UNESCO Chair in Data Privacy,
Dept. of Computer Engineering and Mathematics,
Av. Països Catalans 26, E-43007 Tarragona, Catalonia
{josep.domingo,qianhong.wu}@urv.cat

Abstract. Vehicular *ad hoc* networks (VANETs) will improve traffic safety and efficiency provided that car-to-car communication stays trustworthy. Therefore, it is crucial to ensure that the information conveyed by vehicle-generated messages is reliable. A sensible option is to request that the content of a message originated by a certain vehicle be endorsed by nearby peer vehicles. However, neither message generation nor message endorsement should entail any privacy loss on the part of vehicles co-operating in it. This chapter surveys the available solutions to this security-privacy tension and analyzes their limitations. A new privacy-preserving system is sketched which guarantees message authentication through both *a priori* and *a posteriori* countermeasures.

Keywords: Vehicular *ad hoc* networks, Privacy, Trust, Car-to-car messages.

1   Introduction

According to recent technology forecasts [5], vehicles will be equipped with radio interfaces in the near future and vehicle-to-vehicle (V2V) communications will be available in vehicles by 2011. The IEEE 802.11p task group is working on the Dedicated Short Range Communications (DSRC) standard which aims at enhancing the 802.11 protocol to support wireless data communications for vehicles and the road-side infrastructure [39]. Car manufacturers and telecommunication industry gear up to equip each car with devices known as On-Board Units (OBUs) that allow vehicles to communicate with each other, as well as to supply Road-Side Units (RSUs) to improve safety, traffic efficiency, driver assistance, and transportation regulation. The RSUs are expected to be located in the critical points of the road, such as traffic lights at road intersections. The OBUs and RSUs form a self-organized network called a vehicular *ad hoc* network (VANET), emerging as the first commercial instantiation of the mobile *ad hoc* networking (MANET) technology.

VANETs allow nodes including vehicles or road-side infrastructure units to communicate with each other over single or multiple hops. In other words, nodes will act both as end points and routers. Vehicular networking protocols allow vehicles to broadcast messages to other vehicles in the vicinity. It is suggested

C. Bettini et al. (Eds.): Privacy in Location-Based Applications, LNCS 5599, pp. 173–189, 2009.

that each vehicle periodically send messages over a single hop every 300ms within a distance of 10s travel time (which means a distance range between 10m and 300m)[35]. This mechanism can be used to improve safety and optimize traffic. However, malicious vehicles can also make use of this mechanism by sending fraudulent messages for their own profit or just to jeopardize the traffic system. Hence, the system must be designed to ensure that the transmission comes from a trusted source and has not been tampered with since transmission.

Another critical concern in VANETs is the privacy or anonymity of the driver (or the vehicle, for that matter). As noted in [10], a lot can be inferred about the driver if the whereabouts and the driving pattern of a car can be tracked. It is indeed possible for attackers to trace vehicles by using cameras or physical tracking, but such physical attacks can only trace specific targets and are much more expensive than monitoring the communication in VANETs. Hence, most studies focus on thwarting the latter attacks.

1.1   VANET Architecture

Since in VANETs vehicles periodically report their state information including their location-related information to other vehicles, it is natural to require the vehicles to be equipped with hardware allowing position information collection. Such type of hardware includes GPS or DGPS receivers, which additionally allow clock synchronization. These positioning systems are increasingly available in high-end vehicles. Sensors are another type of devices enabling vehicles to collect their state information including speed, temperature, direction, etc. These devices turn vehicles into potential information sources in view of improving traffic safety and efficiency.

Among the vehicle onboard equipment, some hardware modules are required to guarantee that the information collected has not been altered when sending it to other vehicles [34]. Typically, there are two modules, namely the Event Data Recorder (EDR) and the Tamper-Proof Device (TPD). The EDR only provides tamper-proof storage while the TPD also possesses cryptographic processing capabilities. The EDR is responsible for recording the critical data of the vehicle (such as position, speed, time, etc.) during emergency events, similar to the black box of an airplane. These data will help in accident reconstruction and liability attribution. Hence, if some investigation is later carried out, these messages can be extracted and used as evidences. EDRs are already installed in many road vehicles, e.g. trucks. TPD provides the ability to verify and sign messages. Compared with general CPUs, a TPD not only provides the ability for processing, but it also provides hardware protection so that it cannot be easily penetrated by anyone who is not authorized to do so. The TPD stores all the cryptographic material and performs cryptographic operations, especially signing and verifying safety messages. By binding secret cryptographic keys with a given vehicle, the TPD guarantees the accountability property as long as it remains inside the vehicle. However, TPDs suffer from a high cost for mass deployment and can currently only be expected in high-end vehicles.

VANETs can be viewed as a special kind of mobile *ad hoc* networks. VANET-enabled vehicles are equipped with radio interface for the purpose of forming short-range wireless *ad hoc* networks. In the United States, the Federal Communications Commission has allocated a licensed frequency band of about 75MHz in the 5.8/5.9GHz band specifically for VANETs, usually referred to as DSRC (Dedicated Short Range Communications) [39]. Similar bands have been allocated in Japan and Europe. The MAC layer protocol is either a modified version of 802.11 WLAN or the 3G protocol extended for decentralized access. The original 802.11 protocol is not suitable for VANETs due to the high mobility and highly dynamic topology. A modified version referred to as 802.11p is being developed by the IEEE group for VANETs. The 3G protocol is designed for centralized cellular networks and there have been efforts to enhance it with TDMA- and CDMA-based MAC protocols for decentralized access [27].

A difference between VANETs and generic mobile *ad hoc* networks is that centralized authorities are expected in most VANETs. A conventional transportation regulation system (without VANETs) may involve vehicle manufacturers, a transportation regulation office, the traffic police, and judges. Hence, as suggested in existing schemes [20,26], it is reasonable to assume that those conventional entities have their corresponding electronic counterparts in a VANET. Such centralized authorities are responsible for enrolling vehicles, validating their identities, and issuing electronic certificates to vehicles. The authorities also take care of regular (*e.g.*, annual) health checks of vehicles. In case of serious traffic accidents, they may be involved in collecting electronic witnesses, reconstructing accidents, and tracing drivers. By collecting vehicular communications, some authorities may play a role in optimizing traffic and relieving congestions.

In addition to the above centralized authorities, road-side infrastructure can be expected in vehicular *ad hoc* networks. These distributed road-side units are very useful in collecting vehicular communications and optimizing traffic, distributing and relaying safety-related messages, enrolling vehicles from other VANETs and so on. Depending on the designated roles of the road-side infrastructure, the existing proposals assume very diverse numbers and distributions of road-side units. Some proposals require base stations to be distributed evenly throughout the whole road network, others only at intersections, and others at region borders. Due to cost considerations, especially at the beginning it is unrealistic to require vehicles to always have wireless access to road-side base stations.

1.2   Potential Applications

VANETs have various potential applications including collision avoidance, accident investigation, driving assistance, traffic optimization, traffic regulation, vehicle-based infotainment and so on. Basically, the applications fall into three categories.

Traffic safety related applications. These applications are the main thrust behind VANETs. There are tens of thousands of deaths each year and hundreds of thousands of people get injured in traffic accidents all over the world. Safety related messages from a road-side unit to a vehicle could warn a driver when she/he enters an intersection. V2V communications can save many lives and prevent injuries. Some of the worst traffic accidents are due to many vehicles rear-ending each other after a single accident at the front of the line suddenly halts traffic. In this scenario, if a vehicle broadcasts a message about sudden braking to its neighbor vehicles and other receivers relay the message further, more drivers far behind will get an alarm signal before they see the accident and such type of serious traffic accidents could be avoided. VANETs can also provide driving assistance, *e.g.* violation warning, turn conflict warning, curve warning, lane merging warning etc., to avoid traffic accidents. Many of the accidents come from the lack of co-operation between drivers. By giving more information about the possible conflicts, many life-endangering accidents can be averted. Furthermore, given that vehicle state information and vehicular communications are accountable, VANETs can help in accident reconstruction and witness collection so that the injured and sacrificed can be compensated fairly in case of casualties.

Traffic optimization. With the increasing number of vehicles, people are experiencing more and more traffic delays during the rush hours. VANETs can greatly reduce traffic delays in several ways. Firstly, vehicles could serve as information source and transmit the traffic condition information for the vehicular network. Then transportation agencies could utilize this information to guide vehicles. This will finally relieve traffic congestion. Secondly, vehicles can also work as information collectors and collect data about weather or road surface conditions, construction zones, highway or rail intersections, emergency vehicle signal preemption, etc., and relay those data to other vehicles. Thirdly, the driving assistance provided by VANETs can also improve traffic efficiency. With that assistance, drivers can enjoy smooth and easy driving by avoiding possible conflicts. Finally, VANETs allow transportation administration authorities to manage vehicles electronically (*e.g.* speed control, permits, etc.), which is much more efficient than traditional manual administration.

Value-added services in VANETs. Since vehicles usually have sufficient computational capacity and power supply, complex protocols can be implemented to provide advanced services in VANETs. By implementing advanced electronic payment protocols [3] in VANETs, one can achieve the convenient and desirable goal of passing a toll collection station without having to reduce speed, wait in line, look for some coins and so on. As GPS systems have become available in many vehicles, it is also possible to realize location-based services in VANETs, for instance, finding the closest fuel station, restaurant, hotel, etc. Other kind of services include infotainment, vehicle-based electronic commerce and so on. All these services lead to a more comfortable driving experience and an easier life for drivers, although they are not the main purpose when designing VANETs.

1.3   Plan of the Rest of This Chapter

Section 2 describes the characteristics of the VANETs assumed in the subsequent sections. Section 3 reviews the countermeasures proposed in the literature to obtain secure and privacy-preserving VANETs. Section 4 discusses how to combine *a priori* and *a posteriori* countermeasures in order to overcome the shortcomings of proposals in the literature. Section 5 is a conclusion.

2   Characteristics of VANETs

In this section, we describe the VANET environment in which our scheme is assumed to be implemented.

2.1   Entities in a VANET

Typically, the following entities are present in a VANET:

- **Semi-trusted parties.** A conventional transportation regulation system (without VANETs) may involve vehicle manufacturers, a transportation regulation office, the traffic police, and judges. Hence, as suggested in existing schemes, it is reasonable to assume that those conventional entities have their corresponding electronic counterparts in a VANET. They all have respective secret/public key pairs. These public keys may be embedded into OBUs which are assumed to be tamper-proof. Unlike previous schemes, we adopt a weaker trust assumption that the parties cannot access the private keys of vehicles. A weaker assumption implies more robust systems against vehicle control by, say, organized criminals (for whom it will be harder to access the vehicle's private key).
- **Vehicles.** A VANET mainly consists of vehicles which periodically produce safety-related messages. OBUs and vehicles can be physically bound by embedding an OBU into each vehicle. The owner of a vehicle can also be bound to an OBU by some unique information such as a PIN or his/her unique biometric features. Ownership of a vehicle might be transferred by erasing existing personal information and recording a new one, along with a physical contract. Although the driver might not be the owner, it is the driver who fully controls the vehicle during driving. Hence, we interchangeably use the terms OBU, vehicle, owner and driver.
- **Infrastructures.** As commonly suggested, we assume that there exist centralized authorities and distributed units in a VANET. The centralized authorities can be implemented with the above semi-trusted parties. For instance, the manufacturers produce a signature to show that vehicle ownership is legally transferred to the buyer. With this signature, the vehicle can register to the administration office, and the police office and judges can co-operatively trace the vehicle. Road-side units are also part of the VANET infrastructure. Some power of the authorities can be distributed to road-side units to avoid a communication bottleneck.

- **Attackers.** Since the main security threats in VANETs are violations of public safety and vehicle privacy, we define an attacker to be an entity who wants to successfully cheat honest vehicles by spreading false information, or compromise the privacy of honest vehicles by monitoring the communications in VANETs. A group of maliciously colluding vehicles can also be viewed as an attacker who fully controls that group of vehicles. Attackers can be either internal (that is, VANET entities) or external. They can also be classified as rational or irrational: a rational attacker follows a rational strategy, in which the cost of attack should not be more than its expected benefit, whereas the strategy of an irrational attacker (*e.g.*, a suicide terrorist) cannot be predicted in those terms. Denial-of-service (DoS) is another class of attacks which has been extensively investigated and will not be dealt with here [1].

2.2   Security Requirements

In order to obtain an implementable system to enhance the trustworthiness in V2V communications by balancing public safety and vehicle privacy, we consider the following three types of security requirements:

- **Threshold authentication.** A message is viewed as trustworthy only after it has been endorsed by at least t vehicles, where t is a threshold. The threshold mechanism is an *a priori* countermeasure that improves the confidence of other vehicles in a message. In itself, the threshold does not stop malicious behavior, but makes it more difficult to succeed. Also, the authentication may provide arguments if such behavior occurs and must later be judged.
- **Anonymity.** There is anonymity if, by monitoring the communication in a VANET, message originators cannot be identified, except perhaps by designated parties. The goal is to protect the privacy of vehicles. Since message authentication requires knowledge of a public identity such as a public key or the licence plate, if no anonymity mechanism was provided, an attacker could easily trace any vehicle by monitoring the VANET communication. This would surely be undesirable for the drivers. However, it is possible for attackers to trace vehicles by using a physical approach, *e.g.*, assisted by a camera. But such physical attacks can only trace specific targets and they are much more expensive than attacks monitoring the communication in VANETs. The anonymity mechanism is intended to disable the latter attacks.
- **Revocability.** Revocability means that, if necessary, designated parties can identify the originator and the endorsers of any doubtable message. The goal here is to balance personal privacy and public safety. If anonymity is realized without any revocability mechanism, an attacker can anonymously broadcast authenticated wrong messages to fool other vehicles without fear of being caught, which may seriously compromise public safety. The revocability mechanism is an *a posteriori* countermeasure intended to fight this impunity situation.

2.3   Operational Features of VANETs

Considering the real world, we can assume semi-trusted parties to serve as centralized authorities for vehicle registration and administration. It is also possible to use some road-side units as distributed administration nodes. These administration units make centralized security infrastructures such as public key infrastructures (PKI) usable in VANETs.

The number of mobile nodes in VANETs can be extremely large. Each vehicle is allowed to broadcast messages to other vehicles in the vicinity. It is suggested that each vehicle periodically send messages over a single hop every 300ms within a range of 10s travel time [35]. This yields a minimum range of 10m and a maximum range of 300m, corresponding to the distance covered in 10s travel time. As a consequence, the vehicles are confronted with a large number of message-signature pairs to be verified. Hence, an authentication mechanism designed for VANETs must allow fast message verification.

Due to the road-bounded topology of vehicular networks, common *ad hoc* routing/forwarding strategies are well suited for data dissemination with minimal modifications. It is a reasonable assumption that the intended communication range of an emergency message be greater than the road width. Therefore, a message relayed along the direction of a road will cover all the road area up to destination. This mechanism can also be extended in case of scenarios with road junctions.

The nodes may move very fast at a relative speed up to 320km per hour (*e.g.* two vehicles driving at 160km/h crossing each other in opposite directions). The duration of the connection between mobile nodes may be very short, *e.g.*, less than three seconds. This implies that message-signature pairs should be short enough to be transmitted before the (very sporadic) communication ends.

Unlike the nodes of other types of MANETs which are limited in power and computation, the computation devices embedded in vehicles can be expected to have substantial computational capacity, storage space and power supply. This is the physical basis on which better security can be provided in VANETs.

3   Countermeasures for Securing VANETs

VANETs can improve traffic safety only if the messages sent by vehicles are trustworthy. Dealing with fraudulent messages is a thorny issue for safety engineers due to the self-organized operation of VANETs. The situation is further deteriorated by the privacy requirements of vehicles since, in a privacy-preserving setting, the message generators, *i.e.* the vehicles, are anonymous and cannot be identified when performing maliciously. A number of schemes have been proposed to reduce fraudulent messages; such proposals fall into two classes, namely *a posteriori* and *a priori*.

3.1   *A Posteriori* Countermeasures

A posteriori countermeasures consist in taking punitive action against vehicles which have been proven to have originated fraudulent messages. To be compatible with privacy preservation, these countermeasures require the presence of a trusted third party able to open the identities of dishonest vehicles. Then the identified vehicles can be removed from the system.

Cryptographic authentication technologies have been extensively exploited to offer *a posteriori* countermeasures. Some proposals use regular digital signatures [2,34,36,37] to enable tracing malicious vehicles. To make this approach work, a public key infrastructure (PKI) is suggested in VANETs [31,34,35,41]. These schemes do not provide any methods for certificate revocation. Although issues about revocation were discussed in [7,30,37], no complete solution was provided. To address the above problem in VANETs, [36] proposes three CA revocation protocols: Revocation using Compressed Certificate Revocation Lists (RC2RL), Revocation of the Tamper-Proof Device (RTPD), and Distributed Revocation Protocol (DRP). The RC2RL protocol employs a compression technique to reduce the overhead of distribution of the certificate revocation list. Instead of checking the CA status, The RTPD proposal removes revoked certificates from their corresponding certificate stores in the vehicles. To achieve this, they introduce a tamper-proof device as a vehicle key and certificate management tool. Unlike RC2RL and RTPD, a distributed CA revocation approach is suggested in DRP to determine the status of a certificate. In this case, each vehicle is equipped with an attacker detection system, which enables a vehicle to identify any compromised peer.

When multiple vehicles observe the same driving environment, to endorse the generated message they need to authenticate the same/similar message. This raises the issue of authentication of aggregated data. In [33], the authors propose ways to authenticate identical messages. Another way to deal with authentication of aggregated data is suggested in [32]. This proposal can also handle messages that are similar but not identical, and expects nodes receiving multiple messages with similar information to summarize the information in them using only syntactic aggregation.

A critical issue posed by vehicular message authentication is driver's privacy. Since the public key used to verify the authenticated messages can be linked to specific users, attackers can trace vehicles by observing vehicular communications. Hence, mechanisms must be adopted to guarantee vehicle/driver privacy when vehicles authenticate messages. Along this research line, there are two main approaches: pseudonymous mechanisms and group signatures.

In a pseudonymous mechanism, the certificate authorities produce multiple pseudonyms for each vehicle so that attackers cannot trace the vehicles producing signatures in different periods under different pseudonyms, except if the certificate authorities open the identities of the vehicles. The IEEE 1609.2 Draft Standard [22] proposes the distribution of short-lived certificates to enable vehicle privacy. In [36], the authors propose to use a set of anonymous keys that change frequently (say every couple of minutes) depending on the driving speed. Each key can be used only once and it expires after its usage; only one key can be used at a time. These keys are preloaded in the vehicle's tamper-proof device (TPD) for a long duration, *e.g.* until the next yearly checkup; the TPD takes care of all the operations related to key management and usage. Each key is certified by the issuing CA and has a short lifetime (*e.g.*, one specific week of the year). With the help of the CAs, the key can be tracked back to the real identity of the vehicle in case law enforcement necessitates this and only after obtaining a permission from a judge.

Pseudonym mechanisms have been extensively investigated from various aspects. Short-lived certificates are also suggested in [23], mainly from the perspective of how often a node should change a pseudonym and with whom it should communicate. The authors of [38] propose to use a silent period in order to hamper linkability between pseudonyms, or alternatively to create groups of vehicles and restrict vehicles in one group from hearing messages of other groups. In [19] a user can cloak information before sending it, by providing location information at a coarse granularity in terms of time and space. In [4] mix zones are studied to protect location privacy of location-based services. In [15] the integration of pseudonymity into a real VANET communication system is investigated, bringing together different aspects. Challenges include addressing concepts across layers of the protocol stack, issues in geographical routing (location service, forwarding), and cross layer information exchange, as well as problems related to implementation design and performance.

This conditional anonymity of pseudonymous authentication will help determining the liability of drivers in the case of accidents. The downside of this approach is the necessity for generation, delivery, storage, and verification of numerous of CAs for all the keys. To mitigate this heavy overhead, [7] presents an approach to enable vehicle on-board units to generate their own pseudonyms without interacting with the CAs. The mechanism is realized with the help of group signatures. In [20] a novel group signature-based security framework is proposed which relies on tamper-resistant devices (requiring password access) for preventing adversarial attacks on vehicular networks. However, they provide no concrete instantiation or experiment analysis.

In [26], the authors propose a security and privacy preserving protocol for VANETs by integrating the techniques of group signature and identity-based signature. In their proposal, they take into account security and privacy preservation between OBUs, as well as between OBUs and RSUs. In the former aspect, a group signature is employed to secure the communication between OBUs, where messages are anonymously signed by the senders while the identities of the senders can be traced by the trusted authorities if the messages are later found to be doubtable. In the latter aspect, an identity-based signature scheme is used at RSUs to sign each message generated by RSUs to ensure its authenticity. With their approach, the heavy load of certificate management can be greatly reduced.

Hubaux *et al.* [21] take a different perspective of VANET security and focus on privacy and secure positioning issues. They observe the importance of the tradeoff between liability and anonymity and also introduce Electronic License Plates (ELP) that are unique electronic identities for vehicles.

3.2 *A Priori* Countermeasures

VANETs can improve traffic safety and efficiency only if vehicular messages are correct and precise. Despite the security provided by the combination of TPDs with authenticated messages, an attacker could still manage to transmit valid messages containing false data. It is easy for an attacker to launch such an attack. For instance, putting the vehicle temperature sensor in cold water will let the OBUs generate false messages, even if the hardware sensors are tamper-proof. Also, one may note that in some cases the sender of the data may not necessarily be malicious, but his vehicle's sensors may be out of order. To rule out such cases of false data, one needs not only to verify that the sender of the data is legitimate, but also that the data are correct. Therefore some mechanisms for detection of malicious data need to be explored. We refer to such approaches as *a priori* countermeasures which attempt to prevent the generation of erroneous messages in advance.

The application of information-theoretic measures to anomaly detection was previously studied in the literature [12,14,24], but mainly in the context of the wired Internet. Most notably, [24] successfully applied the notion of relative entropy (also known as the Kullback-Leibler distance) to measure the similarity between two datasets.

Douceur [11] observes that the redundancy checks commonly built into distributed systems to mitigate the threats posed by faulty or malicious participants fail when a single adversary can present multiple distinct identities. Douceur proposes the use of *resource testing* to verify the uniqueness of online identities in a distributed computing environment. Unfortunately, this technique may fail in a VANET if an adversary has more resources than a normal node.

Location is a very important information shared in a VANET. The first proposal aimed at verifying the position data sent by vehicles is presented in [25]. In this proposal, the authors define a number of sanity checks that any vehicle can perform locally on the position information it receives from neighboring vehicles. All position information received by a vehicle is stored for some time period. This is used to perform the checks, the results of which are weighted in order to compute a trust metric for each neighboring vehicle.

A more general proposal that handles both detection and correction of malicious data is given in [17]. The authors assume that the simplest explanation of some inconsistency in the received information is most probably the correct one. The proposal works as follows. Each vehicle maintains a model of the VANET, containing information about the actual physical state of the network, based on the messages it has received. If a new message is received, the vehicle tries to incorporate the information contained therein in its existing model. If this renders the model inconsistent, then the vehicle searches for the minimal set of malicious vehicles and messages that, if removed from the model, would make the model valid again. These vehicles and messages are removed from the model, and the process continues. This approach is expected to be fully developed and combined with the other security mechanisms.

Observing the heavy overhead incurred by the above protocols to correct erroneous messages, some new proposals suggest more efficient threshold mechanisms [9,17,29,31,33] to achieve a similar goal. In these proposals, a message is trusted only if it was endorsed by a number of vehicles in the vicinity. This approach is based on the assumption that most users are honest and will not endorse any message containing false data. Another implicit assumption is the usual common sense that, the more people endorse a message, the more trustworthy it is. Among these schemes, the proposals in [9] may be the most efficient while enabling anonymity of message originators by exploiting secret sharing techniques. But their scheme does not provide anonymity revocability, which may not suit some applications in which anonymity must be revoked "for the prevention, investigation, detection and prosecution of serious criminal offences" [13].

3.3   Discussion on Existing Countermeasures

Unfortunately, neither *a posteriori* nor *a priori* countermeasures suffice on their own to secure VANETs. By taking strict punitive action, *a posteriori* countermeasures can protect against rational attackers producing bogus messages to obtain benefits or pranks. However, they are ineffective against irrational attackers such as terrorists. Even for rational attackers, damage has already occurred when punitive action is taken. It seems that *a priori* countermeasures function better in this case because they prevent damage beforehand by letting the vehicles trust only messages endorsed by a certain number of vehicles. However, although the underlying assumption that there is a majority of honest vehicles in VANETs generally holds, it cannot be guaranteed that a number of malicious vehicles greater than or equal to the threshold will never be present at specific locations. For example, this is likely to happen if some criminal organization undertakes to divert traffic from a certain area by broadcasting messages informing that a road is barred. Furthermore, for convenience of implementation, existing schemes use an even stronger assumption that the number of honest vehicles in all cases should be at least a preset threshold. But such a universally valid threshold does not exist in practice. Indeed, the threshold should somehow take the traffic density and the message scope into account: a low density of vehicles calls for a lower threshold, whereas a high density and a message relevant to the entire traffic of a city requires a sufficiently high threshold.

The situation is aggravated by the anonymity technologies used in some proposals. A system preserves anonymity when it does not require the identity of its users to be disclosed. Without anonymity, attackers can trace all the vehicles by monitoring the communication in VANETs, which in turn can enable the attackers to mount serious attacks against specific targets. Hence, anonymity is a critical concern in VANETs. However, anonymity can also weaken *a posteriori* and *a priori* countermeasures. Indeed, attackers can send fraudulent messages without fear of being caught, due to anonymity; as a result, no punitive action can be taken against them. Furthermore, some proposals provide strong anonymity, *i.e.* unlinkability. Unlinkability implies that a verifier cannot distinguish whether two signatures come from the same vehicle or two vehicles. This feature may enable malicious vehicles to mount the so-called Sybil attack: a vehicle generates a fraudulent message and then endorses the message herself by computing on it as many signatures as required by the threshold in use; since signatures are unlinkable, no one can find out that all of them come from the same vehicle. Hence, elegantly designed protocols are required to secure VANETs when incorporating anonymity. It must be noted that, among those threshold-based systems cited above which provide *a priori* protection and anonymity, [9] is the only one resistant to the Sybil attack: in that system, vehicles belong to groups, and vehicles in a group share keys (which provides vehicle anonymity because vehicles in a group are interchangeable as far as signing goes); however, for a message to be validated, endorsements from a number of different groups are needed, so a single vehicle cannot get a message sufficiently endorsed.

4   Towards a Combination of *a Priori* and *a Posteriori* Countermeasures

Our focus is to devise a context-aware threshold authentication framework with conditional privacy in VANETs, equipped with the following properties:

- It should support a threshold authentication mechanism in the sense that a vehicle can verify whether a received message has been endorsed by at least t vehicles. The threshold can be preset or dynamically changed according to the VANET context.
- It is privacy-preserving. An attacker cannot trace the vehicles who broadcast message-signature pairs. The attacker cannot tell whether the messages are endorsed by the same vehicle or not. This property prevents attackers from identifying vehicles by collecting and mining data.
- It allows revoking anonymity when necessary. As mentioned above, without an anonymity mechanism, t malicious vehicles can anonymously endorse a bogus message to cheat other vehicles. For example, a bang of criminals can divert traffic from their target area by broadcasting a message pretending that the road leading to that area is blocked by snow.

4.1   Message-Linkable Group Signatures

Group signatures have been investigated for many years [8,18]. In a group signature scheme, each group member can anonymously sign messages on behalf of the group. However, a group manager[1] can open the identity of the author of any group signature in case of dispute. Most existing group signatures provide unlinkability in the sense that no efficient algorithm can tell whether two group signatures are generated by the same group member, even if the two signatures are on the same message. Linkable group signatures [28] are a variant of group

---

[1] In most existing schemes, the group manager is responsible for both enrolling members and tracing signers, but some authors suggest to separate these two roles to improve security [6].

signatures. In a linkable group signature, it is easy to identify the group signatures produced by an identical signer, even if the signer is anonymous. This feature is desirable in e-voting systems where each voter can anonymously vote only once.

Group signatures are useful for securing VANETs but they are vulnerable to the Sybil attack because of unlinkability. Linkable group signatures can thwart the Sybil attack but are not compatible with vehicle privacy due to the linkability of signer identities, *i.e.* the various message endorsements signed by a certain vehicle can be linked. Hence, a more sophisticated notion of linkability is required in group signatures for VANETs. Motivated by this observation, we present a new primitive referred to as message-linkable group signatures (MLGS).

An MLGS scheme has the same security properties as regular group signatures except that, given two signatures on *the same message*, one can easily decide whether the two signatures are generated by the same member or by two different members, but the originator(s) stay(s) anonymous. Specifically, a message-linkable signature is an interactive protocol between a register manager, a tracing manager, a set of group members and a set of verifiers. It consists of the following polynomial-time algorithms:

- **Setup**: It is a probabilistic setup algorithm which, on input a security parameter λ, outputs the public system parameters denoted by π, including a description of the system.
- **GKGen**: It is a probabilistic group key generation algorithm which, on input the system parameters π, outputs the public-private key pairs of the register manager and the tracing manager.
- **MKGen**: It is a probabilistic member key generation algorithm which, on input the system parameters π, outputs the public-private key pairs of group members.
- **Join**: It is an interactive protocol between group members, the register manager and the tracing manager. The output of a group member is a group certificate. The output of the register manager is a list of registered group members. The output of the tracing manager is some secret tracing information to trace group signatures.
- **GSign**: It is a probabilistic algorithm which, on input the system parameters π, a message m, a private group member key and the corresponding group certificate, outputs a group signature σ of m.
- **GVerify**: It is a deterministic algorithm which, on input the system parameters π, a message m, a group signature σ and the public key of the register manager, outputs a bit 1 or 0 to represent whether σ is valid or not.
- **GTrace**: It is a deterministic algorithm which, on input the system parameters π, a message m, a *valid* group signature σ and the secret tracing information of the trace manager, outputs the identity of the group member who generated σ.

A secure MLGS scheme must be correct, unforgeable, anonymous, traceable and message-linkable. These properties are defined as follows:

- **Correctness.** It states that GVerify always outputs 1 if all parties honestly follow the MLGS scheme.
- **Unforgeability.** An MLGS scheme is unforgeable if any polynomial-time user who has not registered to the group has only a probability negligible in λ to produce a valid group signature.
- **Anonymity.** An MLGS scheme is anonymous if, given a valid message-signature pair from one of two group members, any polynomial-time attacker has only probability $0.5 + \varepsilon$ of guessing the correct originator of the message-signature pair, where ε is negligible in λ.
- **Traceability.** An MLGS scheme is traceable if any polynomial-time attacker has only a negligible probability in λ to produce a valid group signature such that the output of GTrace is not the identity of the group signature originator.
- **Message-linkability.** An MLGS scheme is message-linkable if there exists a deterministic polynomial-time algorithm which takes as input a message m and two valid group signatures σ_1 and σ_2 on m, and outputs a bit 1 or 0 to represent whether or not the two signatures were generated by the same group member.

4.2 A New Solution Based on Message-Linkable Group Signatures

Based on MLGS, we propose a general framework for threshold authentication with revocable anonymity in VANETs. In this framework, each vehicle registers to a vehicle administration office serving as a group registration manager. When t vehicles wish to endorse some message, they can independently generate an MLGS signature on that message. After validating t MLGS signatures on the message, the verifying vehicle is convinced by the authenticated message. However, if later the message is found incorrect, the police office as well as judges (serving as the tracing manager) can trace the t cheating signers. Here, we assume that an honest signer never needs to sign the same message twice. This assumption is workable by embedding a time-stamp in each message, as suggested in most authentication schemes for VANETs, if the OBU of a vehicle senses the same situation at different times.

From the security properties of MLGS schemes, it is clear that the above framework satisfies the required properties of threshold-variable authentication, anonymity and revocability in VANETs. If $t - 1$ vehicles produce t signatures on the same message, then there exists a group member who has been involved in generating at least two signatures. Such an impersonation can be easily identified since the MLGS scheme is message-linkable. The construction is asymptotically optimal in complexity as the overhead is $O(t)$ in both computation and communication, regardless of the group scale. Hence, the above framework is very suitable for threshold authentication in VANETs.

5 Conclusion and Future Work

In this chapter, we have briefly reviewed the state of the art in VANETs. We have described their architecture and some of their potential applications, especially car-to-car information sharing in view of increasing traffic safety. We have justified why VANETs should be secure *and* preserve the driver's privacy. Security and privacy countermeasures proposed in the literature have been reviewed. In order to overcome the limitations of existing proposals, we have presented a framework combining both *a priori* and *a posteriori* countermeasures. The new framework offers a better balance between public safety and driver privacy in VANETs.

There are also other aspects that should receive more attention in the future. We need a more insightful consideration of the relationship between location privacy and anonymity. Anonymity mechanisms make it hard for attackers to link vehicles at specific locations with their identities. However, the location itself can leak information on vehicle identities. Also, content-based security in VANETs should be studied. Messages in VANETs contain much information about driving patterns. It is possible for attackers to extract much private information by collecting and mining vehicular communications. Finally, application-oriented security is also an open-ended line of work: more and more types of applications will appear in VANETs in the future, each bringing its own new security concerns.

Acknowledgments and Disclaimer

This work was partly supported by the Spanish Government through projects CONSOLIDER INGENIO 2010 CSD2007-0004 "ARES" and TSI2007-65406-C03-01 "E-AEGIS". The first author is partly supported as an ICREA-Acadèmia researcher by the Government of Catalonia. The authors are with the UNESCO Chair in Data Privacy, but their views do not necessarily reflect the position of UNESCO nor commit that organization.

References

1. Aad, I., Hubaux, J.-P., Knightly, R.: Denial of service resilience in ad hoc networks. In: Proceedings of ACM MobiCom, Philadelphia, PA, USA (September 2004)
2. Armknecht, F., Festag, A., Westhoff, D., Zeng, K.: Cross-layer privacy enhancement and non-repudiation in vehicular communication. In: 4th Workshop on Mobile Ad-Hoc Networks (WMAN), Bern, Switzerland (March 2007)
3. Au, M.H., Wu, Q., Susilo, W., Mu, Y.: Compact E-cash from bounded accumulator. In: Abe, M. (ed.) CT-RSA 2007. LNCS, vol. 4377, pp. 178–195. Springer, Heidelberg (2007)
4. Beresford, A., Stajano, F.: Mix Zones: User Privacy in Locationaware Services. In: Proc. of PerSec 2004, pp. 127–131 (March 2004)
5. Blau, J.: Car talk. IEEE Spectrum 45(10), 16 (2008)
6. Boyen, X., Waters, B.: Compact group signatures without random oracles. In: Vaudenay, S. (ed.) EUROCRYPT 2006. LNCS, vol. 4004, pp. 427–444. Springer, Heidelberg (2006)
7. Calandriello, G., Papadimitratos, P., Lioy, A., Hubaux, J.-P.: Efficient and robust pseudonymous authentication in VANET. In: Proceedings of the 4th ACM International Workshop on Vehicular Ad Hoc Networks-VANET 2007, pp. 19–28 (2007)

8. Chaum, D., van Heyst, E.: Group signatures. In: Davies, D.W. (ed.) EUROCRYPT 1991. LNCS, vol. 547, pp. 257–265. Springer, Heidelberg (1991)

9. Daza, V., Domingo-Ferrer, J., Sebe, F., Viejo, A.: Trustworthy privacy-preserving car-generated announcements in vehicular ad hoc networks. IEEE Transactions on Vehicular Technology 58(4), 1876–1886 (2009)

10. Dötzer, F.: Privacy issues in vehicular ad hoc networks. In: Danezis, G., Martin, D. (eds.) PET 2005. LNCS, vol. 3856, pp. 197–209. Springer, Heidelberg (2006)

11. Douceur, J.: The Sybil attack. In: Druschel, P., Kaashoek, M.F., Rowstron, A. (eds.) IPTPS 2002. LNCS, vol. 2429, pp. 251–260. Springer, Heidelberg (2002)

12. Eiland, E., Liebrock, L.: An application of information theory to intrusion detection. In: Proceedings of IWIA 2006 (2006)

13. European Parliament. Legislative resolution on the proposal for a directive of the European Parliament and of the Council on the retention of data processed in connection with the provision of public electronic communication services and amending Directive 2002/58/EC (COM(2005)0438 C6-0293/2005 2005/0182(COD)) (2005)

14. Feinstein, L., Schnackenberg, D., Balupari, R., Kindred, D.: Statistical approaches to DDoS attack detection and response. In: Proceedings of the DARPA Information Survivability Conference and Exposition (2003)

15. Fonseca, E., Festag, A., Baldessari, R., Aguiar, R.-L.: Support of anonymity in VANETs - Putting pseudonymity into practice. In: IEEE Wireless Communications and Networking Conference-WCNC 2007 (2007)

16. Gamage, C., Gras, B., Tanenbaum, A.S.: An identity-based ring signature scheme with enhanced privacy. In: Proceedings of the IEEE SecureComm Conference, pp. 1–5 (2006)

17. Golle, P., Greene, D., Staddon, J.: Detecting and correcting malicious data in VANETs. In: Proceedings of the 1st ACM International Workshop on Vehicular Ad Hoc Networks, pp. 29–37 (2004)

18. Groth, J.: Fully anonymous group signatures without random oracles. In: Kurosawa, K. (ed.) ASIACRYPT 2007. LNCS, vol. 4833, pp. 164–180. Springer, Heidelberg (2007)

19. Gruteser, M., Grunwald, D.: Anonymous usage of location-based services through spatial and temporal cloaking. In: Proc. of MobiSys 2003, pp. 31–42 (2003)

20. Guo, J., Baugh, J.P., Wang, S.: A group signature based secure and privacy-preserving vehicular communication framework. In: Mobile Networking for Vehicular Environments, pp. 103–108 (2007)

21. Hubaux, J.-P., Capkun, S., Luo, J.: The security and privacy of smart vehicles. IEEE Security and Privacy Magazine 2(3), 49–55 (2004)

22. IEEE P1609.2 Version 1 - Standard for Wireless Access in Vehicular Environments - Security Services for Applications and Management Messages (2006)

23. Jakobsson, M., Wetzel, S.: Efficient attribute authentication with applications to ad hoc networks. In: Proceedings of the 1st ACM International Workshop on Vehicular Ad Hoc Networks - VANET 2004 (2004)

24. Lee, W., Xiang, D.: Information-theoretic measures for anomaly detection. In: Proceedings of the IEEE Symposium on Security and Privacy (2001)

25. Leinmüller, T., Maihöfer, C., Schoch, E., Kargl, F.: Improved security in geographic ad hoc routing through autonomous position verification. In: Proceedings of the 3rd ACM International Workshop on Vehicular Vehicular Ad Hoc Networks - VANET 2006, pp. 57–66 (2006)

26. Lin, X., Sun, X., Ho, P.-H., Shen, X.: GSIS: A secure and privacy preserving protocol for vehicular communications. IEEE Transactions on Vehicular Technology 56(6), 3442–3456 (2007)
27. Luo, J., Hubaux, J.-P.: A survey of inter-vehicle communication. Technical Report IC/2004/24, EPFL, Lausanne, Switzerland (2004)
28. Nakanishi, T., Fujiwara, T., Watanabe, H.: A linkable group signature and its application to secret voting. Transactions of Information Processing Society of Japan 40(7), 3085–3096 (1999)
29. Ostermaier, B., Dötzer, F., Strassberger, M.: Enhancing the security of local danger warnings in VANETs - A simulative analysis of voting schemes. In: Proceedings of the Second International Conference on Availability, Reliability and Security, pp. 422–431 (2007)
30. Papadimitratos, P., Buttyan, L., Hubaux, J.-P., Kargl, F., Kung, A., Raya, M.: Architecture for secure and private vehicular communications. In: ITST 2007, Sophia Antipolis, France (2007)
31. Parno, B., Perrig, A.: Challenges in securing vehicular networks. In: Proceedings of the ACM Workshop on Hot Topics in Networks (2005)
32. Picconi, F., Ravi, N., Gruteser, M., Iftode, L.: Probabilistic validation of aggregated data in vehicular ad-hoc networks. In: Proceedings of the 3rd International Workshop on Vehicular Ad hoc Networks - VANET 2006, pp. 76–85 (2006)
33. Raya, M., Aziz, A., Hubaux, J.-P.: Efficient secure aggregation in VANETs. In: Proceedings of the 3rd International Workshop on Vehicular Ad hoc Networks - VANET 2006, pp. 67–75 (2006)
34. Raya, M., Hubaux, J.-P.: Securing vehicular ad hoc networks. Journal of Computer Security (special issue on Security of Ad Hoc and Sensor Networks) 15(1), 39–68 (2007)
35. Raya, M., Hubaux, J.-P.: The security of vehicular ad hoc networks. In: 3rd ACM Workshop on Security of Ad hoc and Sensor Networks-SASN 2005, pp. 11–21 (2005)
36. Raya, M., Papadimitratos, P., Aad, I., Jungels, D., Hubaux, J.-P.: Eviction of misbehaving and faulty nodes in vehicular networks. IEEE Journal on Selected Areas in Communications 25(8), 1557–1568 (2007)
37. Raya, M., Papadimitratos, P., Hubaux, J.-P.: Securing vehicular communications. IEEE Wireless Communications Magazine 13(5), 8–15 (2006)
38. Sampigethaya, K., Huang, L., Li, M., Poovendran, R., Matsuura, K., Sezaki, K.: CARAVAN: Providing Location Privacy for VANET. In: Proc. of ESCAR 2005 (November 2005)
39. U.S. Department of Transportation, National Highway Traffic Safety Administration, Vehicle Safety Communications Project, Final Report (April 2006), http://www-nrd.nhtsa.dot.gov/pdf/nrd-12/060419-0843/PDFTOC.htm
40. Wu, Q., Domingo-Ferrer, J.: Balanced trustworthiness, safety and privacy in vehicle-to-vehicle Communications, Manuscript (2008)
41. Zarki, M.E., Mehrotra, S., Tsudik, G., Venkatasubramanian, N.: Security issues in a future vehicular network. In: Proceedings of European Wireless 2002 (2002)

Privacy Preserving Publication
of Moving Object Data

Francesco Bonchi

Yahoo! Research
Avinguda Diagonal 177, Barcelona, Spain
bonchi@yahoo-inc.com

Abstract. The increasing availability of space-time trajectories left by location-aware devices is expected to enable novel classes of applications where the discovery of consumable, concise, and actionable knowledge is the key step. However, the analysis of mobility data is a critic task by the privacy point of view: in fact, the peculiar nature of location data might enable intrusive inferences in the life of the individuals whose data is analyzed. It is thus important to develop privacy-preserving techniques for the publication and the analysis of mobility data.

This chapter provides a brief survey of the research on *anonymity preserving data publishing of moving objects databases*.

While only few papers so far have tackled the problem of anonymity in the off-line case of publication of a moving objects database, rather large body of work has been developed for anonymity on relational data on one side, and for location privacy in the on-line, dynamic context of *location based services* (LBS), on the other side. In this chapter we first briefly review the basic concepts of k-anonymity on relational data. Then we focus on the body of research about privacy in LBS: we try to identify some useful concepts for our static context, while highlighting the differences, and discussing the inapplicability of some of the LBS solutions to the static case. Next we present in details some of the papers that recently have attacked the problem of moving objects anonymization in the static context. We discuss in details the problems addressed and the solutions proposed, highlighting merits and limits of each work, as well as the various problems still open.

1 Introduction

Recent years have witnessed the pervasiveness of location-aware devices, e.g., GSM mobile phones, GPS-enabled PDAs, location sensors, and active RFID tags. This new capability of localizing moving objects and persons enables a wide spectrum of possible novel applications that were simply infeasible only few years ago. Those applications can be roughly divided in two large groups:

on-line: such as monitoring the moving objects, real-time analysis of their motion patterns, and development of location-based services;

C. Bettini et al. (Eds.): Privacy in Location-Based Applications, LNCS 5599, pp. 190–215, 2009.
© Springer-Verlag Berlin Heidelberg 2009

off-line: such as the collection of the traces left by these moving objects, the off-line analysis of these traces with the aim of extracting behavioral knowledge in support of, e.g., mobility-related decision making processes, sustainable mobility, and intelligent transportation systems [1].

The latter scenario, which is the focus of this paper, is rapidly gaining a great deal of attention as witnessed by the amount spatio-temporal data mining techniques that have been developed in the last years [2,3,4,5,6,7]. Such techniques may be used, for instance, by governments to measure day-by-day variability in mobility behavior[1], by researchers to study masses mobility behavior[2] or by companies to track employees and maximize their efficiency[3]. Clearly, in these applications privacy is a concern, since location data enables intrusive inferences, which may reveal habits, social customs, religious and sexual preferences of individuals, and can be used for unauthorized advertisement and user profiling.

More concretely, consider a *traffic management application* on a city road network, where the trajectories of vehicles equipped with GPS devices are recorded and analyzed by the city municipality traffic management office. This is not unrealistic: in the context of the GeoPKDD[4] project, we received a dataset of this kind from the city of Milan (Italy). Indeed, many citizens accept to equip their car with GPS devices, because this way they obtain a substantial discount on the mandatory car insurance. Suppose now that the office owning the data is going to outsource the data mining analysis to an external laboratory. In a naïve tentative of preserving anonymity, the car identifiers are not disclosed but instead replaced with pseudonyms. However, as shown in [8], such operation is insufficient to guarantee anonymity, since location represents a property that in some circumstances can lead to the identification of the individual. For example, if Bob is known to follow almost every working morning the same route, it is very likely that the starting point is Bob's home and the ending point is his working place. Joining this information with some telephone directories we can easily link Bob's trajectory to Bob's identity.

As another example, a recent paper by Terrovitis and Mamoulis analyzes the case of a company in Hong Kong, called Octopus[5], that collects daily trajectory data of Hong Kong residents who use Octopus smart RFID card [9]. The data could be disclosed to a third external party, e.g., for *reach of poster analysis* in Hong Kong. The reach of a poster defines the percentage of people who have at least one contact with a given poster (or a posters network) within a specified period of time. The reach allows to determine the optimal duration of some advertisement and to tune the formation of poster networks[6].

---

[1] See http://www.fhwa.dot.gov/ohim/gps/

[2] See the projects of the sens*able* City Lab http://senseable3.mit.edu/

[3] http://www.denverpost.com/headlines/ci_4800440

[4] GeoPKDD: Geographic Privacy-aware Knowledge Discovery and Delivery, EU project IST-6FP-014915, webpage: http://www.geopkdd.eu/

[5] http://www.octopuscards.com

[6] See for instance the Swiss Poster Research http://www.spr-plus.ch/

As Terrovitis and Mamoulis [9] pointed out, when a person, say Alice, uses her Octopus card to pay at different convenience stores that belong to the same chain (e.g., 7-Eleven), by collecting her transaction history in all these stores, the company can construct a subset of her complete trajectory. If this constructed trajectory uniquely identifies Alice, then by matching it with the published trajectory database (this is usually called *"linkage attack"*), even though the users identifiers may be removed, Alice still can be re-identified, as can the other locations that she visited.

The two abovementioned examples regard cases of data released to a third party, but privacy issues do not arise only when the data must be published. As it happens for any other kind of data, collecting, storing and analyzing person-specific mobility data is subject to privacy regulations[7,8]. Therefore it is important to develop concepts and methods for collecting, storing and publishing spatio-temporal data in a way that preserves privacy of the individuals.

The research problem surveyed in this paper can be stated as *anonymity preserving data publishing of moving objects databases*. More formally, we consider a static *moving object database* (MOD) $D = \{O_1, ..., O_n\}$ that correspond to n individuals, a set of m discrete time points $T = \{t_1, ..., t_m\}$, and a function $\mathcal{T} : D \times T \rightarrow \mathcal{R}^2$, that specifies, for each object O and a time t, its position at time t. The function $\mathcal{T}$ is called the trajectory. Indeed, $\mathcal{T}(O_i)$ denotes the trajectory of object O_i, i.e., $\mathcal{T}(O_i) = \{(x_i^1, y_i^1, t_1), ..., (x_i^m, y_i^m, t_m)\}$ is O_i's trajectory, with (x_i^j, y_i^j) representing the position of O_i at time t_j. The problem is how to transform D in such a way that it satisfy some form of anonymity, while most of its original utility is maintained in the transformed database D^*.

Introduced by Samarati and Sweeney [10,11,12], the concept of k-anonymity has established, also thanks to its simplicity, as the *de facto* standard solution to prevent linkage attacks in de-identified relational databases. The idea behind k-anonymity can be described as "hiding in the crowd", as it requires that each release of data must be such that each individual is indistinguishable from at least $k - 1$ other individuals. In the classical k-anonymity framework the attributes are partitioned into *quasi-identifiers* (i.e., a set of attributes whose values can be linked to external information to reidentify the individual), and *sensitive attributes* (publicly unknown, which we want to keep private). In order to provide k-anonymity, the values of the quasi-identifiers are generalized to be less specific so that there are at least k individuals in the same group, who have the same (generalized) quasi-identifer values. Although it has been shown that the k-anonymity model presents some flaws and limitations [13], and that finding an optimal k-anonymization is NP-hard [14,15], it remains a fundamental model of privacy with practical relevance.

Unfortunately (and quite obviously), the concept of k-anonymity can not be borrowed from relational databases as it is, because in the case of moving objects much more complexity is brought in by the peculiar nature of spatio-temporal

---

[7] http://www.cdt.org/privacy/eudirective/EUDirective.html
[8] http://www.dataprotection.ie/documents/legal/6aiii.htm

data. In the rest of this paper we will discuss this complexity, and we will analyze in details the few papers that have attacked this problem in the last two years.

Paper Content and Structure

So far only few papers have tackled the problem of anonymity in the off-line case of publication of a moving objects database. Instead, a lot of work has been done for anonymity on relational data, and for location privacy in the on-line, dynamic context of *location based services* (LBS). In Section 2 we review the basic concepts of k-anonymity on relational data, while in Section 3 we review the body of research on privacy in LBS. We try to identify some useful concepts for our static context, while highlighting the differences, and discussing the inapplicability of some of the LBS solutions to the static case. In particular, we recall the interesting concepts of *location based quasi-identifier* and *historical k-anonymity* introduced by Bettini *et al.* in [8].

Then we focus on four papers (all very recent) that, to the best of our knowledge, are the unique that have attacked the problem of MOD anonymization so far. Since one key concept is that of quasi-identifier, we use this concept to partition the methods in two groups: methods assuming no quasi-identifier and thus anonymizing the trajectories in their whole [16,17] (Section 4) and methods assuming some form of quasi-identifer [9,18] (Section 5).

Finally, in Section 6 we draw some conclusions and discuss the open research problems.

2   Relational Data Anonymity

Samarati and Sweeney showed that the simple de-anonymization of individual sources does not guarantee protection when sources are cross-examined: a sensitive medical record, for instance, can be uniquely linked to a *named* voter record in a publicly available voter list through some shared attributes. The objective of k-anonymity is to eliminate such opportunities of inferring private information through cross linkage.

The traditional k-anonymity framework [10,11,12,19] focuses on relational tables: the basic assumptions are that the table to be anonymized contains entity-specific information, that each tuple in the table corresponds uniquely to an individual, and that attributes are divided in *quasi-identifier* (i.e., a set of attributes whose values in combination can be linked to external information to reidentify the individual to whom the information refers); and *sensitive attributes* (publicly unknown and that we want to keep secret).

According to this approach, the data holder has the duty of identifying all possible attributes in the private information that can be found in other public databases, i.e., the attributes that could be exploited by a malicious adversary by means of cross linkage (the quasi-identifier).

Once the quasi-identifier is known, the "anonymization" of the database takes place: the data is transformed in such a way that, for every combination of values of the quasi-identifier in the sanitized data, there are at least k records that share

Table 1. (a) example medical table, (b) a 2-anonymous version of table (a), (c) an alternative 2-anonymous version of table (a), and (d) a 2-anonymous version of table (a) by full-domain generalization. These example tables are borrowed from [20].

Job	Birth	Postcode	Illness
Cat1	1975	4350	HIV
Cat1	1955	4350	HIV
Cat1	1955	5432	flu
Cat1	1955	5432	fever
Cat2	1975	4350	flu
Cat2	1975	4350	fever

(a)

Job	Birth	Postcode	Illness
Cat1	*	4350	HIV
Cat1	*	4350	HIV
Cat1	1955	5432	flu
Cat1	1955	5432	fever
Cat2	1975	4350	flu
Cat2	1975	4350	fever

(b)

Job	Birth	Postcode	Illness
*	1975	4350	HIV
*	*	4350	HIV
Cat1	1955	5432	flu
Cat1	1955	5432	fever
*	*	4350	flu
*	1975	4350	fever

(c)

Job	Birth	Postcode	Illness
*	*	4350	HIV
*	*	4350	HIV
*	*	5432	flu
*	*	5432	fever
*	*	*	flu
*	*	4350	fever

(d)

those values. One equivalence class of records sharing the same quasi-identifier values is usually called *anonymity set* or *anonymization group*.

The anonymization is usually obtained by (i) generalization of attributes (the ones forming the quasi-identifier), and (ii), when not avoidable, suppression of tuples [12].

An example medical data table is given in Table 1(a). Attributes `job`, `birth` and `postcode` form the quasi-identifier. Two unique patient records (corresponding to the first two rows) may be re-identified easily since their combinations of the attributes forming the quasi-identifier are unique. The table is generalized as a 2-anonymous table in Table 1(b): here the two patient records are indistinguishable w.r.t. the quasi-identifier and thus are less likely to be re-identified by means of cross linkage.

In the literature of k-anonymity, there are two main models. One model is *global recoding* [12,19,20,21,22] while the other is *local recoding* [12,14,20]. A common assumption is that each attribute has a corresponding conceptual hierarchy or taxonomy. Generalization replaces lower level domain values with higher level domain values. A lower level domain in the hierarchy provides more details and maintains more of the original information than a higher level domain. In global recoding, all values of an attribute come from the same domain level in the hierarchy. For example, all values in `birth` date are in years, or all are in both months and years. One advantage is that an anonymous view has uniform domains, but the price to pay is higher information loss. For example, a global recoding of Table 1(a) is in Table 1(d), but it clearly suffers from overkilling generalization.

With local recoding, values may be generalized to different levels in the domain. For example, Table 2 is a 2-anonymous table by local recoding. In fact one can say that local recoding is a more general model and global recoding is a special case of local recoding. Note that, in the example, known values are replaced by unknown values (*), indicating maximum generalization, or total loss of information.

As discussed by Domingo-Ferrer and Torra in [23] the methods based on generalization and suppression suffer of various drawbacks. To overcome some of these limitations the use of *microaggregation* for k-anonymity has also been proposed [23]. Microaggregation is a concept originating from the statistical disclosure control (SDC) research community. In particular, under the name microaggregation goes a family of perturbative SDC methods that have been developed both for continuous and categorical data, and that do not require a hierarchy [23,24,25,26]. Whatever the data type, microaggregation can be operationally defined in terms of the following two steps:

- *Partition:* the set of original records is partitioned into several clusters in such a way that records in the same cluster are similar to each other and so that the number of records in each cluster is at least k.
- *Aggregation:* An aggregation operator (for example, the mean for continuous data or the median for categorical data) is computed for each cluster and is used to replace the original records. In other words, each record in a cluster is replaced by the clusters prototype.

This approach, even if under different names, e.g. *k-member clustering for k-anonymity*, has been investigated in [27], and then extended in [28,29] to deal with attributes that have a hierarchical structure. Usually, after a clustering step what is released is the centroid of each cluster together with the cardinality of the cluster.

Another similar approach is introduced by Aggarwal and Yu in [30]. *Condensation* is a perturbation-like approach which aims at preserving the inter-attribute correlations of data. It starts by partitioning the original data into clusters of exactly k elements, then it regenerates, for each group, a set of k fake elements that approximately preserves the distribution and covariance of the original group. The record regeneration algorithm tries to preserve the eigenvector and eigenvalues of each group. The general idea is that valid data mining models (in particular, classification models) can be built from the reconstructed data without significant loss of accuracy. Condensation has been applied by the same authors also to sequences [31].

Some limitations of the k-anonymity model, with consequent proposals for improvement, have emerged in the literature. One first drawback is that the difference between quasi-identifiers and sensitive attributes may be sometimes vague, leading to a large number of quasi-identifiers. This problem has been studied in [32], where Aggarwal analyzes the scalability of distortion w.r.t. the number of dimensions used in the k-anonymization process, showing that for sparse data the usability of the anonymized data could sensibly decrease. A possible solution based on k-anonymity parameterized w.r.t. a given public dataset has been proposed by Atzori in [33]. Xiao and Tao in [34] propose a decomposition of quasi-identifiers and sensitive data into independently shared databases. Kifer and Gehrke [35] suggest to share anonymized marginals (statistical models such as density estimates of the original table) instead of the private table.

Another drawback of simple k-anonymity is that it may not protect sensitive values when their entropy (diversity) is low: this is the case in which a sensitive value is too frequent in a set of tuples with same quasi-identifier values after k-anonymization [13,20,36,37]. Consider Table 1(b): although it satisfies 2-anonymity property, it does not protect two patients sensitive information, HIV infection. We may not be able to distinguish the two individuals for the first two tuples, but we can derive the fact that both of them are HIV infectious. Suppose one of them is the mayor, we can then confirm that the mayor has contracted HIV. Surely, this is an undesirable outcome. Note that this is a problem because the other individual whose generalized identifying attributes are the same as the mayor also has HIV. Table 3 is an appropriate solution. Since (*,1975,4350) is linked to multiple diseases (i.e. HIV and fever) and (*,*,4350) is also linked to multiple diseases (i.e. HIV and flu), it protects individual identifications and hides the implication.

Regardless of these limitations, k-anonymity remains a widely accepted model both in scientific literature and in privacy related legislation, and in recent years a large research effort has been devoted to develop algorithms for k-anonymity (see for instance [22,38,39] just to cite some of the most relevant ones).

3   Anonymity and Location Based Services

As witnessed by the other chapters in this volume, most of the existing work about anonymity of spatio-temporal moving points has been developed in the context of *location based services* (LBS). In this context a trusted server is usually in charge of handling users' requests and passing them to the service providers, and the general goal is to provide the service on-the-fly without threatening the anonymity of the user that is requiring the service.

This is the main difference with our setting where we have a static database of moving objects and we want to publish it in such a way that the anonymity of the individuals is preserved, but also the *quality* of the data is kept high. On the contrary, in the LBS context the aim is to provide the service without learning user's exact position, and ideally the data might also be forgotten once that the service has been provided. In other terms, in our context anonymity is *off-line* and *data-centric*, while in the LBS context is a sort of *on-line* and *service-centric* anonymity. A solution to the first problem is not, in general, a solution to the second (and viceversa), and both problems are important. However, although in our context the focus is on the quality of the data, while in LBS is on the quality of the service, it should be noted that both concepts of quality are intrinsically geared on the level of precision with which positions are represented.

In the following we review some of the proposals for privacy in LBS trying to identify useful concepts for our static context, while discussing why some of the LBS solutions are not suitable for our purposes.

The concept of *location k-anonymity* for location based services was first introduced in [40] Gruteser and Grunwald, and later extended by Gedik and Liu in [41] to deal with different values of k for different requests. The underlying idea is that a message sent from a user is k-anonymous when it is indistinguishable from the spatial and temporal information of at least $k - 1$ other messages sent from different users. The proposed solution is based on a spatial subdivision in areas, and on *delaying the request* as long as the number of users in the specified area does not reach k. The work in [41] instead of using the same k for all messages, allows each message to specify an independent anonymity value and the *maximum spatial* and *temporal tolerance resolutions* it can tolerate based on its privacy requirements. The proposed algorithm tries to identify the smallest spatial area and time interval for each message, such that there exist at least $k - 1$ other messages from different users, with the same spatial and temporal dimensions. Domingo-Ferrer applied the idea of microaggregation for k-anonymity in location-based services [42].

Kido *et al.* [43] propose a privacy system that takes into account only the spatial dimension: the area in which location anonymity is evaluated is divided into several regions, and position information is delimited by the region it belongs to. Anonymity is required in two different ways: the first, called *ubiquity*, requires that a user visits at least k regions; the second, called *congestion*, requires the number of users in a region to be at least k. High ubiquity guarantees the location anonymity of every user, while high congestion guarantees location anonymity of local users in a specified region.

In [44] Beresford and Stajano introduce the concept of *mix zones*. A mix zone is an area where the location based service providers can not trace users' movements. When a user enters a mix zone, the service provider does not receive the real identity of the user but a pseudonym that changes whenever the user enters a new mix zone. In this way, the identities of users entering a mix zone in the same time period are mixed. A similar classification of areas, named *sensitivity map* is introduced by Gruteser and Liu in [45]: locations are classified as either *sensitive* or *insensitive*, and three algorithms that hide users' positions in sensitive areas are proposed.

Contrary to the notions of mixed zones and sensitivity maps, the approach introduced by Bettini *et al.* in [8] is geared on the concept of *location based quasi-identifier*, i.e., a spatio-temporal pattern that can uniquely identify one individual. More specifically, a location based quasi-identifier (LBQID) is a sequence of spatio-temporal constraints (i.e., a spatial area and a time interval) plus a recurrence formula.

Example 1 (Borrowed from [8]). A user may consider the trip from the condominium where he lives to the building where he works every morning and the trip back in the afternoon as an LBQID if observed by the same service provider for at least 3 weekdays in the same week, and for at least 2 weeks. This LBQID may be represented by the spatio-temporal pattern:

$$\langle AreaCondominium\,[7am, 8am],\ AreaOfficeBldg\,[8am, 9am],$$
$$AreaOfficeBldg\,[4pm, 6pm],\ AreaCondominium\,[5pm, 7pm]\rangle$$
$$Recurrence: 3.Weekdays * 2.Weeks$$

where the various areas such as *AreaCondominium* identify sets of points in bidirectional space possibly by a pair of intervals $[x_1, x_2][y_1, y_2]$.

Where and how LBQID are defined for each single user, is an interesting open problem not addressed in [8], and we will discuss it again later in this chapter. In [8] Bettini *et al.* simply state that the derivation process of those spatio-temporal patterns to be used as LBQID will have to be based on statistical analysis of the data about users movement history: if a certain pattern turns out to be very common for many users, it is unlikely to be useful for identifying any one of them. Since in the LBS framework there is the trusted server which stores, or at least has access to, historical trajectory data, Bettini *et al.* argue that the trusted server is probably a good candidate to offer tools for LBQID identification. However, the selection of candidate patterns may also possibly be guided directly by the users.

The goal of introducing the concept of LBQID is that of ensuring that no sensitive data is released from the trusted server receiving the requests, to a service provider, when the data can be personally identified through a LBQID. On the technical level, considering the language proposed in Example 1, a timed state automata [46] may be used for each LBQID and each user, advancing the state of the automata when the actual location of the user at the request time is within the area specified by one of the current states, and the temporal constraints are satisfied.

In [8] the concept of *historical k-anonymity* is also introduced. Given the set of requests issued by a certain user (that corresponds to a trajectory of a moving object in our terminology), it satisfies historical k-anonymity if there exist $k - 1$ *personal histories of locations* (i.e., trajectories in our terminology) belonging to $k - 1$ different users such that they are *location-time consistent* (i.e., undistinguishable). What the framework in [8] aims at, is to make sure that if a set of requests from a user matches a location based quasi-identifier then it satisfies historical k-anonymity. The idea is that if a service provider can successfully track the requests of a user through all the elements of a LBQID, then there would be at least $k - 1$ other users whose personal history of locations is consistent with these requests, and thus may have issued those requests. This is achieved by generalization in space and time, essentially by increasing the uncertainty about the real user location and time of request. Generalization is performed by an algorithm that tries to preserve historical k-anonymity of the set of requests that have matched the current partial LBQID. If for a particular request, generalization fails, (i.e., historical k-anonymity is violated), the system will try to unlink future requests from the previous ones by means of the mixed zones technique [44] that we discussed earlier.

Although defined in the LBS context this work is very relevant to the problem surveyed in this chapter, i.e., anonymity in a static database of moving objects.

In other terms, a set of trajectories that satisfies historical k-anonymity may be considered safe also in our data-publishing context. The main difference, however, is the fact that they consider data point (requests) continuously arriving, and thus they provide *on-line anonymity*. More concretely, the anonymization group of an individual is chosen once and for all, at the time this is needed, i.e., when his trajectory matches a LBQID. This means that the $k - 1$ moving objects passing closer to the point of the request are selected to form the anonymization group regardless of what they will do later, as this information is not yet available. Instead in our context the information about the whole history of trajectories is available, thus we must select anonymization groups considering the trajectories in their whole. This is the main reason why the solution proposed in [8] does not seem suitable for our static data-publishing context.

As discussed at the beginning of this section the difference is that in the LBS context the emphasis is on providing the service, while in our context the emphasis is on the quality maintained in the anonymized database. As another example consider again the concept of *mix zones* previously described: it is a solution for LBS, since it provides some sort of anonymity for the users while still allowing the service to be provided correctly; but it is not a solution for data publishing, since the quality of the data is completely destroyed. Just think about a data mining analysis (e.g., finding hot routes [6], frequent mobility patterns [47] clustering trajectories [4], etc.) on a dataset "anonymized" by means of the mixed-zone technique: the results would simply be unreliable.

Summarizing, in this section we have reviewed techniques for anonymity in the LBS context, and we have discussed the differences between this context and our context: why the anonymity solutions proposed in the former are not suitable in the latter. However, many techniques developed in the LBS research community, such as location perturbation, spatio-temporal cloaking (i.e., generalization), and the personalized privacy-profile of users, should be taken in consideration while devising solutions for anonymity preserving data publishing of MODs. Also the definitions of historical k-anonymity and LBQID might be borrowed in the context of data publishing: how to do it is a challenging open problem not addressed in [8] nor in other work.

4   Methods Based on Clustering and Perturbation

The problem of defining a concept of quasi-identifier for a database of moving objects, that is both realistic and actionable, is not an easy task. On the one hand, it is the nature itself of spatio-temporal information to make the use of quasi-identifiers unlikely, or at least, challenging. The natural spatial and temporal dependence of consecutive points in a trajectory, avoids from identifying a precise set of locations, or a particular set of timestamps to be the quasi-identifier. Moreover, as argued in [18], unlike in relational microdata, where every tuple has the same set of quasi-identifier attributes, in mobility data we can not assume to have the same quasi-identifier for all the individuals. It is very likely that various moving objects have different quasi-identifiers and this should be taken into account in modeling adversarial knowledge. But as shown in [18] allowing different quasi-identifiers for different objects creates many challenges: the main one being that the anonymization groups may not be disjoint.

A different order of challenges regarding the definition of a concept of quasi-identifier for MODs, arises at the practical level. It is not clear from where and how the quasi-identifier for each single user should be defined. Both [8] and [18] argue that the quasi-identifiers may be provided directly by the users when they subscribe to the service, or be part of the users personalized settings, or they may be found by means of statistical data analysis or data mining. However, the problem of spatio-temporal quasi-identifier definition in the real-world is an open issue.

Given the aforementioned challenges, Abul *et al.* [16], and Nergiz *et al.* [17], have tackled the problem of anonymity of MODs without considering any concept of quasi-identifier, thus anonymizing trajectories as a whole. This may be considered as implicitly and conservatively assuming that the adversary may identify each user in any location at any time. Since the proposed techniques provide protection under this very conservative setting, they also provide protection under less powerful adversary.

In this setting the output of anonymization can only be a set of anonymization groups each one containing identical, or at least very similar, sets of trajectories and having size at least k. From the above consideration, it follows that a natural approach is to tackle the anonymity problem by means of clustering, more precisely *microaggregation* or *k-member clustering*, i.e., clustering with the constraint on the minimum population of each cluster.

The first work tackling the anonymization of trajectories as a constrained clustering problem is [16]. In that paper Abul *et al.* propose a novel concept of k-anonymity based on co-localization that exploits the inherent uncertainty of the moving object's whereabouts. Due to sampling and positioning systems (e.g., GPS) imprecision, the trajectory of a moving object is not simply a polyline in a three-dimensional space, instead it is a cylindrical volume, where its radius δ represents the possible location imprecision: we know that the trajectory of the moving object is within this cylinder, but we do not know exactly where. A graphical representation of an uncertain trajectory is reported in Figure 1(a).

If another trajectory moves within the cylinder (or uncertainty area) of the given trajectory, then the two trajectory are indistinguishable from each other (or in other terms, they're a *possible motion curve* of each other). This leads to the definition of (k, δ)-anonymity for moving objects databases. More formally, given an anonymity threshold k, Abul *et al.* define a (k, δ)-*anonymity set* as a set of at least k trajectories that are co-localized w.r.t. δ. Then they show that a set of trajectories S, with $|S| \geq k$, is a (k, δ)-anonymity set if and only if there exists a trajectory τ_c such that all the trajectories in S are possible motion curves of τ_c within an uncertainty radius of $\delta/2$. Given a (k, δ)-anonymity set S, the trajectory τ_c is obtained by taking, for each $t \in [t_1, t_n]$, the point (x, y) that is the center of the minimum bounding circle of all the points at time t of all trajectories in S. Therefore, an anonymity set of trajectories can be bounded

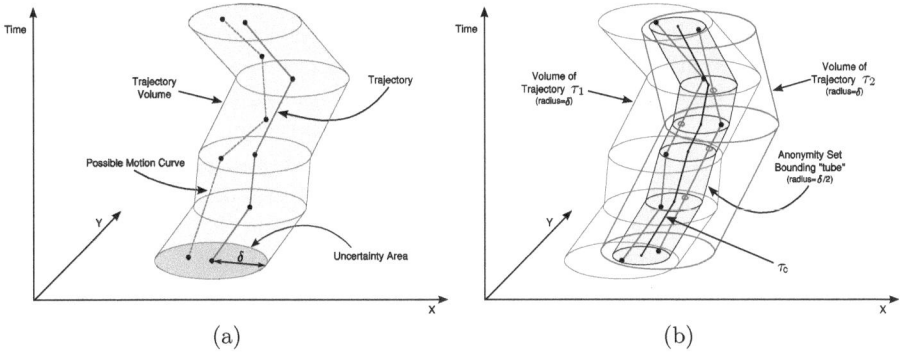

Fig. 1. (a) an uncertain trajectory: uncertainty area, trajectory volume and possible motion curve. (b) an anonymity set formed by two co-localized trajectories, their respective uncertainty volumes, and the central cylindrical volume of radius $\delta/2$ that contains both trajectories.

by a cylindrical volume of radius $\delta/2$. In Figure 1(b), we graphically represent this property.

The problem of (k, δ)-anonymizing a database of trajectories of moving objects requires to transform a MOD D in D^* such such that for each trajectory $\tau \in D^*$ it exists a (k, δ)-anonymity set $S \subseteq D^*$, $\tau \in S$, and the distortion between D and D^* is minimized.

In [16] a two-step method, based on clustering and perturbation, is devised to achieve (k, δ)-anonymity. In particular, as perturbation method is chosen *space translation*: i.e., slightly moving some observations in space. A suitable measure of the information distortion introduced by space translation is defined, and the problem of achieving (k, δ)-anonymity by space translation with minimum distortion is proven to be NP-hard.

In the first clustering step, the MOD D is partitioned in groups of trajectories, each group having size in the interval $[k, 2k - 1]$. After having tried a large variety of clustering methods for trajectories under the k-member constraint, Abul *et al.* chose a simple greedy method as the best trade-off between efficiency and quality of the results. The resulting method, named $\mathcal{NWA}$ ($\mathcal{N}$ever $\mathcal{W}$alk $\mathcal{A}$lone), is further enhanced with ad-hoc preprocessing and outlier removal. In fact it is claimed by the authors (but also by other previous work, e.g., [29]), that outlier detection and removal might be a very important technique in clustering-based anonymization schemes: the overall quality of the anonymized database can benefit by the removal of few outlying trajectories.

The pre-processing step aims at partitioning the input database into larger equivalence classes w.r.t. time span, i.e. groups containing all the trajectories that have the same starting time and the same ending time. This is needed because $\mathcal{NWA}$ adopts Euclidean distance that can only be defined among trajectories having the same time span: if performed directly on the raw input data this often produces a large number of very small equivalence classes, possibly leading to very low quality anonymization. To overcome this problem, a simple pre-processing method is developed. The method enforces larger equivalence classes at the price of a small information loss. The pre-processing is driven by an integer parameter π: only one timestamp every π can be the starting or ending point of a trajectory. For instance, if the original data was sampled at a frequency of one minute, and $\pi = 60$, all trajectories are pre-processed in such a way that they all start and end at full hours. To do that, the first and the last suitable timestamps occurring in each trajectory are detected, and then all the points of the trajectory that do not lay between them are removed.

The greedy clustering method iteratively selects a pivot trajectory and makes a cluster out of it and of its $k - 1$ unvisited nearest neighbors, starting from a random pivot and choosing next ones as the farthest unvisited trajectories w.r.t. previous pivots. Being simple and extremely efficient, the greedy algorithm allows to iteratively repeat it until clusters satisfying some criteria of compactness are built.

More in details, a compactness constraint is added to the greedy clustering method briefly described above: clusters to be formed must have a radius not larger than a given threshold. When a cluster cannot be created around a new pivot without violating the compactness constraint, the latter is simply *deactivated* — i.e., it will not be used as pivot but, in case, it can be used in the future as member of some other cluster — and the process goes on with the next pivot. When a remaining object cannot be added to any cluster without violating the compactness constraint, it is considered an outlier and it is trashed. This process might lead to solutions with a too large trash, in which case the whole procedure is restarted from scratch relaxing the compactness constraint, reiterating the operation till a clustering with sufficiently small trash is obtained. At the end, the set of clusters obtained is returned as output, thus implicitly discarding the trashed trajectories.

In the second step, each cluster of trajectories is perturbed by means of the minimum spatial translation needed to push all the trajectories within a common uncertainty cylinder, i.e., transforming them in an anonymity set.

Data quality of the anonymized database D^* is assessed both by means of objective measures of information distortion, and by comparing the results of the same spatio-temporal range queries executed on D and D^*. In particular, as of objective measures Abul *et al.* adopt the total information distortion introduced by the spatial translation of points, and *discernibility*. Introduced in [38], discernibility is a simple measure of the data quality of the anonymized dataset based on the size of each anonymity set. Given a clustering $\mathcal{P} = \{p_1, \ldots, p_n\}$ of $\mathcal{D}$, where p_n represents the trash bin, the discernibility metric is defined as: $DM(D^*) = \sum_{i=1}^{n-1} |p_i|^2 + |p_n||D^*|$. Intuitively, discernibility represents the fact that data quality shrinks as more data elements become indistinguishable. The experiments reported in [16] show that discernibility is strongly influenced by the number of removed trajectories, and it does not provide any information about the amount of distortion introduced, thus resulting not much suitable for the cases of trajectory anonymization.

Abul *et al.* also report experiments on range query distortion, adopting the model of *spatio-temporal range queries with uncertainty* of [48]. In that work it is defined a set of six (Boolean) predicates that give a qualitative description of the relative position of a moving object τ with respect to a region R, within a given time interval $[t_b, t_e]$. In particular the condition of interest is $inside(R, \tau)$. Since the location of the object changes continuously, we may ask if such condition is satisfied *sometime* or *always* within $[t_b, t_e]$; moreover, due to the uncertainty, the object may *possibly* satisfy the condition or it may *definitely* do so (here the uncertainty is expressed by the same δ of the anonymization problem). If there exists some possible motion curve which at the time t is inside the region R, there is a possibility that the moving object will be inside R at t. Similarly, if every possible motion curve of the moving object is inside the region R at the time t, then regardless of which one describes the actual objects motion, the object is guaranteed to be inside the region R at time t. Thus, there are two domains of quantification, with two quantifiers in each. In [16], only the two extreme cases are used in the experimentation: namely *Possibly_Sometime_Inside*, corresponding to a double $\exists$, and *Definitely_Always_Inside*, corresponding to a double $\forall$. The query used is the count of the number of objects in D, and for comparison in D^*, satisfying *Possibly_Sometime_Inside* (or *Definitely_Always_Inside*) for some randomly chosen region R and time interval $[t_b, t_e]$ (averaging on a large number of runs). Experimental results show that for a wide range of values of δ and k, the relative error introduced by the method of Abul *et al.* is kept reasonably low. Also the running time is shown to be reasonable even on large MODs.

Inspired by the *condensation* approach [30,31], Nergiz *et al.* [17] tackles the trajectory anonymization problem by means of grouping and reconstruction. In their framework no uncertainty is part of the input of the problem: an anonymity set is defined as a set of size $\geq k$ of *identical* trajectories (this correspond to the case of $\delta = 0$ in the setting of [16]). To make this feasible it is necessary to generalize points in space and time. Since they consider generalization, trajectory is defined as a sequence of 3D spatio-temporal volumes. In other terms each observation, each point in a trajectory is represented by intervals on the three dimensions: $[x_1, x_2]$, $[y_1, y_2]$, and $[t_1, t_2]$.

Therefore, a k-anonymization of a MOD D is definesd as a another MOD D^* such that:

- for every trajectory in D^*, there are at least $k - 1$ other trajectories with exactly the same set of points;
- there is a one to one relation between the trajectories $tr \in D$ and trajectories $tr^* \in D^*$ such that for each point $p_i \in tr^*$ there is a unique $p_j \in tr$ such that $t_i^1 \leq t_j^1$, $t_i^2 \geq t_j^2$, $x_i^1 \leq x_j^1$, $x_i^2 \geq x_j^2$, $y_i^1 \leq y_j^1$ and $y_i^2 \geq y_j^2$.

Given a set of trajectories that are going to be anonymized together, the anonymity set is created by *matching* points, and then by taking the 3D minimum bounding box that contains the matched points. A depiction of the proposed anonymization process is provided in Figure 2.

As clustering strategy Nergiz *et al.* adapt the condensation based grouping algorithm given in [30]. The cost of the *optimal* anonymization is adopted as

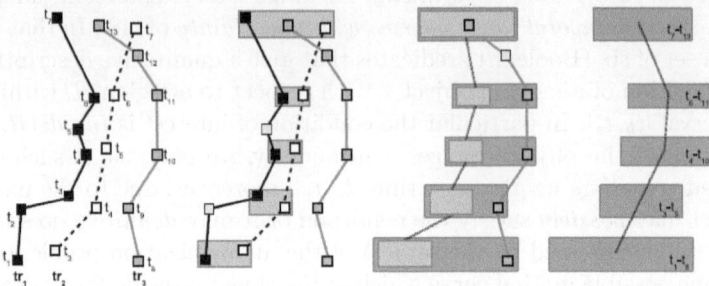

Fig. 2. Anonymization of three trajectories tr_1, tr_2 and tr_3, based on point matching and removal, and spatio-temporal generalization

distance metric between two trajectories. Finding the optimal anonymization of two trajectories is the same as finding the point matching between the two trajectories such that anonymizing the trajectories through the matching minimizes the generalization cost. A similar alignment problem is well studied for strings (where the goal is to find an alignment of strings such that total pairwise edit distance between the strings is minimized) in the context of DNA comparisons. Alignment problem for two trajectories is polynomial and can be solved by using a dynamic programming approach.

The resulting greedy algorithm, named *multi TGA*, at each iteration creates an empty group G, randomly samples one trajectory $tr \in D$, puts tr into G, and initialize the group representative $rep_G = tr$. Next, the closest trajectory $tr' \in TR \setminus G$ to rep_G is selected and added to G, and then rep_G is updated as the bounding box anonymizing rep_G and tr'.

The main drawback of this algorithm is the costly operation of finding the closest trajectory to the group representative. In order to decrease the number of times that such costly operation must be performed, a new algorithm (named *fast TGA*) is introduced: in *fast TGA* all the $k-1$ closest trajectories to the group representative are chosen in one pass. However, another drawback arises, as the challenge now becomes the computation of the optimal anonymization. In fact, while optimal matching between two trajectories is easy, computing the optimal point matching for $n > 2$ trajectories in NP-hard. For tackling this problem Nergiz *et al.* rely on heuristics that have proven to be effective in the string alignment problem.

After providing their generalization-based approach to k-anonymity of trajectories, Nergiz *et al.* discuss some drawbacks of such approach, and suggest that in many cases it might be more suitable to publish a reconstructed MOD, instead of a generalized one. In particular, they claim that generalization suffers from two main shortcomings. Firstly, the use of minimum bounding boxes in anonymization discloses uncontrolled information about exact locations of the points: e.g., in the case of two trajectories, two non-adjacent corners give out the exact locations. Secondly, it is challenging to take full advantage of information

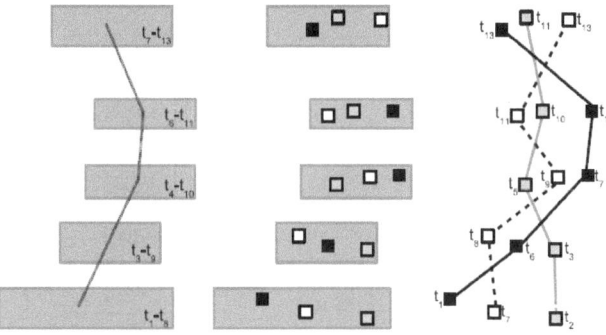

Fig. 3. Example of reconstruction starting from the anonymization of Figure 2

contained in generalized MODs as most data mining and statistical applications work on atomic trajectories.

Therefore Nergiz *et al.* adapt the reconstruction approach [30] and publish reconstructed data rather than data anonymized by means of generalization. An example reconstruction is shown in Figure 3. The output after reconstruction is atomic and suitable for trajectory data mining applications.

For assessing the quality of the resulting anonymization Nergiz *et al.* focus on the utility of the data for mining purposes. In particular, they chose a standard clustering method and compare the results obtained by clustering the original MOD D and its anonymized version D^*. In order to asses the result of clustering, they consider every pair of trajectories and verify whether both are in the same cluster, in the clustering given by D, and whether they are in the same cluster, in the clustering given by D^*. Then they measure accuracy, precision and recall.

5   Methods Based on Quasi-identifier

In this section we review two recent approaches to anonymization of MODs that adopt some concept of quasi-identifier.

The basic assumption of work by Terrovitis and Mamoulis [9] is that the adversaries own portions of the moving objects, and different adversaries owns different parts. The portion of a trajectory known by an adversary may be used to perform a linkage attack if the MOD is published without paying attention to anonymity. The privacy that is required is that, from the data publication, an adversary can not learn anything more than what he already knows.

As motivating example, they analyze the case of a company in Hong Kong called Octopus that collects daily trajectory data of Hong Kong residents who use Octopus smart RFID card. As we discussed in the motivating example in Section 1, when Alice uses her Octopus card to pay at different convenience stores that belong to the same chain (e.g., 7-Eleven), she left a sequence of traces, in some sense, giving away to the company a portion of her own trajectory. If this

Table 2. (a) an example MOD D, and (B) a local MOD D^A (A's knowledge)

t_{id}	trajectory
t_1	$a_1 \to b_1 \to a_2$
t_2	$a_1 \to b_1 \to a_2 \to b_3$
t_3	$a_1 \to b_2 \to a_2$
t_4	$a_1 \to a_2 \to b_2$
t_5	$a_1 \to a_3 \to b_1$
t_6	$a_3 \to b_1$
t_7	$a_3 \to b_2$
t_8	$a_3 \to b_2 \to b_3$

(a)

t_{id}	trajectory
t_1^A	$a_1 \to a_2$
t_2^A	$a_1 \to a_2$
t_3^A	$a_1 \to a_2$
t_4^A	$a_1 \to a_2$
t_5^A	$a_1 \to a_3$
t_6^A	a_3
t_7^A	a_3
t_8^A	a_3

(b)

projection of her trajectory uniquely identifies Alice, then by matching it with the published trajectory database, even though the IDs of users may be removed, Alice still can be re-identified, as can the other locations outside the portion of Alice's trajectory that 7-Eleven already knows.

More formally, Terrovitis and Mamoulis consider trajectories being simple sequences of addresses, corresponding to the places in which the Octopus card is used. Let $\mathcal{P}$ be the domain of all addresses where the Octopus card is a accepted. Since commercial companies might have multiple branches, $\mathcal{P}$ can be partitioned in m disjoint non-empty sets of addresses $\mathcal{P}_1, \mathcal{P}_2, \ldots, \mathcal{P}_m$ such that each set contains all and only the addresses of the different branches of a company. Or in other terms, each adversary i controls a portion of addresses $\mathcal{P}_i$. For each trajectory t in the input MOD D, each adversary i holds a portion (or a projection) t^i. In general, each adversary i holds a local database D^i containing the projections of all $t \in D$ with respect to $\mathcal{P}_i$. The adversary has no knowledge about trajectories having empty projection; therefore, $\mathcal{P}_i$ can be smaller than the database of the publisher. A trajectory may appear multiple times in D and more than one trajectories may have the same projection with respect to $\mathcal{P}_i$. The most important property of a t^i is that adversary i can directly link it to the identities of all persons that pass through it, in its local database (e.g., loyalty program). Consider the example MOD D given in Table 2(a). Each sequence element is a shop address, where the corresponding user did his/her card transactions. Locations are classified according to the possible adversaries. For example, all places denoted by a_i are assumed to also be tracked by company A (e.g., 7-Eleven). Table 2(b) shows the knowledge of A. This knowledge D^A can be combined with D, if D is published, to infer private information. For instance, if D is published, A will know that t_5^A actually corresponds to t_5, since t_5 is the only trajectory that goes through a_1 and a_3, and no other location of company A. Therefore A is 100% sure that the user whose trajectory is t_A^5, visited b_1.

Therefore, the problem tackled by Terrovitis and Mamoulis in [9] can be formulated as follows. Given a MOD D, where each trajectory $t \in D$ is a sequence of values from domain $\mathcal{P}$, construct a transformed database D^*, such that if D^* is public, for all $t \in D$, every adversary i cannot correctly infer any location $\{p_j | p_j \in t \wedge p_j \notin t^i\}$ with probability larger than a given threshold P_{br}.

This problem is similar to the l-diversity problem defined in [13]. The main differences with the problem of privacy preserving publication in the classic relational context, are that in this context quasi-identifiers are variable-length sequences of locations, and that there can be multiple sensitive values (i.e., locations) per trajectory and these values are different from the perspectives of different adversaries. The second difference is that the algorithm which transforms D in D^* must consider linkage attacks to different sensitive values from different adversaries at the same time. One important point is that anonymization is based on the assumption that the data owner is aware of the adversarial knowledge, i.e., which adversary holds which portion of data, or in other terms, the data owner is exactly aware of the partition $\mathcal{P}_1, \mathcal{P}_2, \ldots, \mathcal{P}_m$ of $\mathcal{P}$.

As objective function of utility Terrovitis and Mamoulis adopt the average difference between the original trajectories in D and the published ones in D^*. The method used to sanitize D from the possible linkage attacks is based on the identification, and the consequent suppression, of certain points that causes potential threats. This is done taking under consideration the benefit in terms of privacy and the deviation from the main direction of the trajectory. Since the problem of finding the optimal set of points to delete from D in order to derive a secure D^* and achieve the minimum possible information loss is harder NP-hard, they proposes a greedy algorithm that iteratively suppresses locations, until the privacy constraint is met. The algorithm simulates the attack from any possible adversary, and then solves the identified privacy breaches. The algorithm is empirically evaluated by measuring the effective cost and the number of points suppressed.

While this problem statement fits perfectly the scenario described above, it is not easily adaptable for other cases, e.g., the scenario of mobility data, where a set of cars equipped with GPS move on a road-network, or users with mobile phones move in a city. In this cases it seems less reasonable to assume that the data owners knows which are the spatio-temporal points known by an adversary.

Yarovoy *et al.* [18] in another recent work addresses the problem of privacy-preserving publication of MODs, considering quasi-identifiers, and using spatial generalization as anonymization technique. The authors argue that unlike in relational microdata, where every tuple has the same set of quasi-identifier attributes, in mobility data we can not assume a set of particular locations, or a set of particular timestamps, to be a quasi-identifier for all the individuals. It is very likely that various moving objects have different quasi-identifiers and this should be taken into account in modeling adversarial knowledge.

More precisely, given a MOD $D = \{O_1, ..., O_n\}$ that correspond to n individuals, a set of m discrete time points $T = \{t_1, ..., t_m\}$, and a function $\mathcal{T} : D \times T \rightarrow \mathbb{R}^2$, that specifies, for each object O and a time t, its position at time t, they consider timestamps as attributes with objects' positions forming their values, and they assume quasi identifiers to be sets of timestamps. As said above, a fixed set of timestamps can not be the quasi-identifier for all the moving objects. To capture this, they define the quasi-identifier as a function:

MOB	t_1	t_2
O_1	$(1,2)$	$(5,3)$
O_2	$(2,3)$	$(2,7)$
O_3	$(6,6)$	$(3,6)$

(a)

MOB	t_1	t_2
O_1	$[(1,2),(2,3)]$	$(5,3)$
O_2	$[(1,2),(2,3)]$	$[(2,6),(3,7)]$
O_3	$(6,6)$	$[(2,6),(3,7)]$

(b)

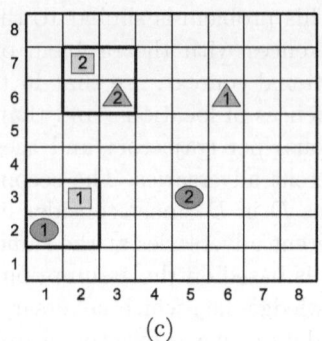

(c)

Fig. 4. Assuming $QID(O_1) = \{t_1\}$, $QID(O_2) = QID(O_3) = \{t_2\}$: (a) original database; (b) a 2-anonymity scheme that is not safe, and (c) its graphical representation

$QID : \{O_1, ..., O_n\} \rightarrow 2^{\{t_1, ..., t_n\}}$. That is, every moving object may potentially have a distinct quasi-identifier.

The main issue in anonymizing MOD is that, due to the fact that different objects may have different QID, anonymization groups associated with different objects may not be disjoint, as illustrated below.

Example 2. Consider the trajectories in Figure 4(a) and illustrated in Figure 4(c). Let $k = 2$ and $QID(O_1) = \{t_1\}$, $QID(O_2) = QID(O_3) = \{t_2\}$. Intuitively the best (w.r.t. information loss) anonymization group for O_1 w.r.t. its QID $\{t_1\}$ is $AS(O_1) = \{O_1, O_2\}$. This is illustrated in Figure 4(c) with a dark rectangle. This means in the anonymized database we assign the region $[(1, 2), (2, 3)]$ to O_1 and O_2 at time t_1. The best anonymization group for O_2 as well as for O_3 w.r.t. their QID $\{t_2\}$ is $\{O_2, O_3\}$. Thus, in the anonymized database, O_2 and O_3 will both be assigned to the common region $[(2, 6), (3, 7)]$ (the second dark rectangle) at time t_2. Clearly, the anonymization groups of O_1 and O_2 overlap.

Due to this fact providing a robust and sound definition of k-anonymity in the case of MOD is challenging, as it will be clarified below. In order to explain why, we first need to introduce some basic definitions.

Given a MOD D, a distorted version of D is any database D^* over the same time points $\{t_1, ..., t_n\}$, where D^* contains one row for every moving object O in D, and either $D^*(O, t) = D(O, t)$ or $D(O, t) \sqsubseteq D^*(O, t)$, where with $\sqsubseteq$ we denote spatial containment among regions. The goal, as usual, is to find a distorted version of the MOD D, denoted by D^*, such that on the one hand, when published, D^* is still useful for analysis purposes, and on the other, a suitable version of k-anonymity is satisfied. The anonymization technique considered is *space generalization*. In the input MOD D, each position is an exact point, but with the application of a grid, each point may be regarded as a cell (as in Figure 4(c)), and generalized points are rectangles made of these cells.

Generalization obviously results in information loss. Yarovoy *et al.* measure information loss as the reduction in the probability with which the position of an object at a given time can be accurately determined. More formally, given a distorted version D^* of a MOD D, the information loss is defined as: $\text{IL}(D, D^*) = \sum_{i=1}^{n} \sum_{j=1}^{m} (1 - 1/area(D^*(O_i, t_j)))$; where $area(D^*(O_i, t_j))$ denotes the area of the region $D^*(O_i, t_j)$. As an example, consider the generalized MOD D^* as in Figure 4(b). The information loss associated with D^* is $2 \times (1 - 1/4) + 2 \times (1 - 1/4) = 3$.

A basic building block for devising any notion of anonymity is a notion of indistinguishability. Let D^* be a distorted version of a MOD D, two moving objects O, O' are indistinguishable in D^* at time t provided that $D^*(O, t) = D^*(O', t)$, i.e., both are assigned to the same region in D^*. The most obvious way of defining k-anonymity is the following: a distorted version D^* of a MOD D satisfies k-anonymity provided that for every moving object O in D, $\exists k - 1$ other distinct moving objects $O_1, ..., O_{k-1}$ in D^*: $\forall t \in QID(O)$, O is indistinguishable from each of $O_1, ..., O_{k-1}$ at time t.

According to this definition the database in Figure 4(b) is 2-anonymous and thus "safe". This obvious definition of k-anonymity still suffers privacy breaches. Indeed, due the fact that anonymization groups may not be disjoint, it is possible that by combining overlapping anonymization groups, some moving objects may be uniquely identified, as explained next. Recall the previous example. There, I_1 and I_2 are in the same anonymization group (i.e., have the same generalized location) at time point t_1 (i.e., the QID of I_1), while I_2 and I_3 are in the same anonymization group at time point t_2 (i.e., the QID of I_2 and I_3). However, when the adversary tries to map the three moving objects O_1, O_2, O_3 to the three individuals I_1, I_2, I_3, with the adversary knowledge of QID values of these three moving objects, he can infer that I_1 must be mapped to either O_1 or O_2, while I_2 (and I_3) should be mapped to either O_2 or O_3. If I_1 is mapped to O_2, we cannot find a consistent assignment for I_2, I_3. As a result, the adversary can conclude that O_1 must map to I_1. Thus, a more sophisticated definition of k-anonymity is needed in order to avoid privacy breaches in the case of moving object databases.

Given the considerations above, Yarovoy et al. [18] define k-anonymity by formalizing the attack described above. In particular they define an attack graph associated with a MOD D and its distorted version D^*, as the bipartite graph G consisting of nodes for every individual I in D (called I-nodes) and nodes for every moving object id O (called O-nodes) in the published database D^*. G contains an edge (I, O) iff $D(O, t) \sqsubseteq D^*(O, t), \forall t \in QID(I)$.

An assignment of individuals to moving objects is consistent provided there exists a perfect matching in the bipartite graph G. Consider the distorted database shown in Figure 4(b): the corresponding attack graph is shown in Figure 5(a). It is obvious that the edge (I_1, O_1) must be a part of every perfect matching. Thus, by constructing the attack graph an attacker may easily conclude that MOB O_1 can be re-identified as I_1.

One of the key shortcomings in the straightforward definition of k-anonymity given above is that while it ensures every I-node corresponding to an individual has at least k neighbors, it does not have any restriction on the degree of the O-nodes. *What if we required that in addition, every O-node must have degree at*

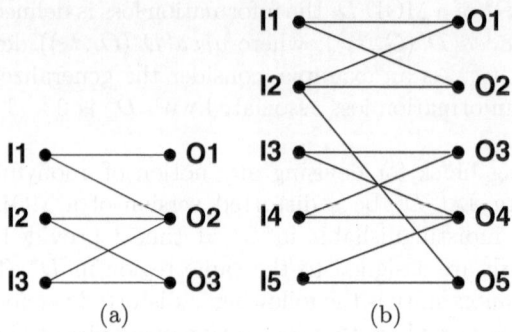

Fig. 5. Attack graphs for different anonymization schemes: (a) for D^* in Figure 4(b); (b) for a hypothetical database D^* satisfying modified definition of k-anonymity

least k? Suppose we say that a distorted database is k-anonymous provided in the corresponding attack graph, every I-node as well as every O-node has degree $\geq k$. Figure 5(b) shows a possible attack graph that satisfies this condition. In this graph, every I-node and every O-node has degree 2 or more. Yet, O_5 can be successfully re-identified as I_5 as follows. Suppose O_5 is instead assigned to I_2, to which it is adjacent as well. Then it is easy to see that no I-node can be assigned to one of O_1, O_2. Thus, *the edge (I_2, O_5) cannot be a part of any perfect matching*. Thus, this edge can be pruned, leaving I_5 as the only I-node to which O_5 can be assigned.

This example is subtler than the previous example and clearly shows the challenges involved in devising a notion of k-anonymity that does not admit privacy breaches.

The attack model is formalized as following. The attacker first constructs an attack graph associated with the published distorted version of D and the known $QIDs$ as described above. Then, he repeats the following operation until there is no change to the graph:

1. Identify an edge e that cannot be part of any perfect matching.
2. Prune the edge e.

Next, he identifies every node O with degree 1. He concludes the (only) edge incident on every such node must be part of every perfect matching. There is a privacy breach if the attacker succeeds in identifying at least one edge that must be part of every perfect matching.

Finally k-anonymity is defined. Let D be a MOD and D^* its distorted version. Let G be the attack graph w.r.t. D, D^*. Then D^* is k-anonymous provided that (i) every I-node in G has degree k or more; and (ii) G is symmetric, i.e., whenever G contains an edge (I_i, O_j), it also contains the edge (I_j, O_i). An immediate observation is that in an attack graph that satisfies the above conditions, every O-node will have degree k or more as well.

Yarovoy *et al.* [18] develop two different algorithms and show that both of them satisfy the above definition of k-anonymity. One main challenge in devising these algorithms arises again from the fact that anonymization groups may not be disjoint: in particular, is overlapping anonymization groups can force the algorithm to revisit earlier generalizations, and possibly re-generalize them with other objects. For computing the anonymity group of a given moving object, both algorithms use a method based on Hilbert index of spatial objects for efficient indexing of trajectories. In the empirical comparison of the two algorithms, Yarovoy *et al.* report statistics on the size of the equivalence classes created in the anonymization, as well as the average information loss introduced. They also report range query distortion similarly to [16].

6 Conclusions and Open Research Issues

We provided an overview of a rather young research effort concerning how to anonymize a moving objects database. While only few papers have been published so far on this problem, much large body of work has been developed for location privacy in the on-line, dynamic context of location based services. We briefly reviewed this body of research trying to clarify why, even if apparently similar, the problem of anonymization becomes deeply different when tackled in a static instead of a dynamic context. However, many techniques developed in the LBS research community, such as location perturbation, spatio-temporal cloaking (i.e., generalization), and the personalized privacy-profile of users, should be taken in consideration while devising solutions for anonymity preserving data publishing of MODs. Also the definitions of *historical k-anonymity* and *location based quasi-identifier* introduced in [8] might be borrowed in the context of data publishing.

We discussed the challenge of deriving quasi-identifiers in the context of mobility data: as argued by some authors, they might be defined by the users themselves, or they might be "learnt" by mining a MOD. Finding a realistic and actionable definition of quasi-identifiers, as well as devising methodology to derive them, are important open problems.

Yarovoy *et al.* [18] argue that, contrarily to the classic relational setting, in MODs quasi-identifiers can only be defined on the individual basis, i.e., each moving object must have his own quasi-identifier. They also show how many computational challenges arise from this assumption. The main one is that the anonymization problem is no longer about finding a partition of objects in disjoint anonymity sets, because due to the different quasi-identifiers, anonymity sets may overlap.

An interesting approach is the one of Terrovitis and Mamoulis [9], that instead of defining quasi-identifiers by the user perspective, consider the linkage attacks that are possible given that different adversaries have knowledge of different parts of the data.

Other authors [16,17] instead do not consider quasi-identifiers and focus on anonymizing trajectories in their whole, by grouping together similar trajectories, and slightly perturbing them, to make them undistinguishable. Even if these approaches avoid the challenges connected to quasi-identifiers, they still face some open problems. One of these is the so-called diversity problem [13] introduced for relational data. In that context it is shown that k-anonymity alone does not put us on the safe side, because although one individual is hidden in a group (thanks to equal values of the quasi-identifier attributes), if the group has not enough diversity of the sensitive attributes then an attacker can still associate one individual to sensitive information. Also in the MOD context, if we are able to know that one individual belong to a group, even if we are not able to identify exactly his trajectory, we can still discover some sensitive information.

Another line of research, not yet started, is about developing ad-hoc anonymization techniques for the intended use of the data: for instance, with respect to a specific spatio-temporal data mining analysis.

A close and interesting research area is the so called *privacy-preserving data mining*, i.e., instead of anonymizing the data for a privacy-aware data publication, the focus of privacy is shifted directly to the analysis methods. Privacy preserving data mining, is an hot and lively research area which has seen the proliferation of many completely different approaches having different objectives, application contexts and using different techniques [49,50,51,52,53]. However, very little work has been done about developing privacy-preserving mining techniques for spatio-temporal and mobility data [54,55,56,57]: as said for the anonymization of MOD, this research topic is rather young and we expect to see many new proposals in the next future.

References

1. Giannotti, F., Pedreschi, D. (eds.): Mobility, Data Mining and Privacy - Geographic Knowledge Discovery. Springer, Heidelberg (2008)
2. Lee, J.G., Han, J., Li, X.: Trajectory outlier detection: A partition-and-detect framework. In: Proc. of the 24th IEEE Int. Conf. on Data Engineering (ICDE 2008) (2008)
3. Lee, J.G., Han, J., Li, X., Gonzalez, H.: raClass: trajectory classification using hierarchical region-based and trajectory-based clustering. In: Proc. of the 34th Int. Conf. on Very Large Databases (VLDB 2008) (2008)
4. Lee, J.G., Han, J., Whang, K.Y.: Trajectory clustering: a partition-and-group framework. In: Proc. of the 2007 ACM SIGMOD Int. Conf. on Management of Data (SIGMOD 2007) (2007)
5. Li, X., Han, J., Kim, S., Gonzalez, H.: Anomaly detection in moving object. In: Intelligence and Security Informatics, Techniques and Applications. Studies in Computational Intelligence, vol. 135. Springer, Heidelberg (2008)
6. Li, X., Han, J., Lee, J.G., Gonzalez, H.: Traffic density-based discovery of hot routes in road networks. In: Papadias, D., Zhang, D., Kollios, G. (eds.) SSTD 2007. LNCS, vol. 4605, pp. 441–459. Springer, Heidelberg (2007)
7. Nanni, M., Pedreschi, D.: Time-focused clustering of trajectories of moving objects. Journal of Intelligent Information Systems 27(3), 267–289 (2006)
8. Bettini, C., Wang, X.S., Jajodia, S.: Protecting Privacy Against Location-Based Personal Identification. In: Jonker, W., Petković, M. (eds.) SDM 2005. LNCS, vol. 3674, pp. 185–199. Springer, Heidelberg (2005)
9. Terrovitis, M., Mamoulis, N.: Privacy preservation in the publication of trajectories. In: Proc. of the 9th Int. Conf. on Mobile Data Management (MDM 2008) (2008)

10. Samarati, P., Sweeney, L.: Generalizing data to provide anonymity when disclosing information (abstract). In: Proc. of the 17th ACM Symp. on Principles of Database Systems (PODS 1998) (1998)
11. Samarati, P., Sweeney, L.: Protecting Privacy when Disclosing Information: k-Anonymity and its Enforcement Through Generalization and Suppresion. In: Proc. of the IEEE Symp. on Research in Security and Privacy, pp. 384–393 (1998)
12. Sweeney, L.: k-anonymity privacy protection using generalization and suppression. International Journal on Uncertainty Fuzziness and Knowledge-based Systems 10(5) (2002)
13. Machanavajjhala, A., Gehrke, J., Kifer, D., Venkitasubramaniam, M.: l-diversity: privacy beyond k-anonymity. In: Proc. of the 22nd IEEE Int. Conf. on Data Engineering (ICDE 2006) (2006)
14. Aggarwal, G., Feder, T., Kenthapadi, K., Motwani, R., Panigrahy, R., Thomas, D., Zhu, A.: Anonymizing tables. In: Eiter, T., Libkin, L. (eds.) ICDT 2005, vol. 3363, pp. 246–258. Springer, Heidelberg (2005)
15. Meyerson, A., Willliams, R.: On the complexity of optimal k-anonymity. In: Proc. of the 23rd ACM Symp. on Principles of Database Systems (PODS 2004) (2004)
16. Abul, O., Bonchi, F., Nanni, M.: $\mathcal{N}$ever $\mathcal{W}$alk $\mathcal{A}$lone: Uncertainty for anonymity in moving objects databases. In: Proc. of the 24nd IEEE Int. Conf. on Data Engineering (ICDE 2008) (2008)
17. Nergiz, E., Atzori, M., Saygin, Y.: Towards trajectory anonymization: a generalization-based approach. In: Proc. of ACM GIS Workshop on Security and Privacy in GIS and LBS (2008)
18. Yarovoy, R., Bonchi, F., Lakshmanan, L.V.S., Wang, W.H.: Anonymizing moving objects: How to hide a MOB in a crowd? In: Proc. of the 12th Int. Conf. on Extending Database Technology (EDBT 2009) (2009)
19. Samarati, P.: Protecting respondents' identities in microdata release. IEEE Trans. Knowl. Data Eng. 13(6), 1010–1027 (2001)
20. Wong, R.C.W., Li, J., Fu, A.W.C., Wang, K. (α, k)-anonymity: an enhanced k-anonymity model for privacy preserving data publishing. In: Proc. of the 12th ACM SIGKDD Int. Conf. on Knowledge Discovery and Data Mining (KDD 2006) (2006)
21. Fung, B.C.M., Wang, K., Yu, P.S.: Top-down specialization for information and privacy preservation. In: Proc. of the 21st IEEE Int. Conf. on Data Engineering (ICDE 2005) (2005)
22. LeFevre, K., DeWitt, D.J., Ramakrishnan, R.: Incognito: Efficient full-domain k-anonymity. In: Proc. of the 2005 ACM SIGMOD Int. Conf. on Management of Data (SIGMOD 2005) (2005)
23. Domingo-Ferrer, J., Torra, V.: Ordinal, continuous and heterogeneous -anonymity through microaggregation. Data Min. Knowl. Discov. 11(2), 195–212 (2005)
24. Defays, D., Nanopoulos, P.: Panels of enterprises and confidentiality: the small aggregates method. In: Proc. of 92 Symposium on Design and Analysis of Longitudinal Surveys, Ottawa, Statistics Canada, pp. 195–204 (1993)
25. Domingo-Ferrer, J., Mateo-Sanz, J.M.: Practical data-oriented microaggregation for statistical disclosure control. IEEE Trans. Knowl. Data Eng. 14(1), 189–201 (2002)
26. Torra, V.: Microaggregation for categorical variables: A median based approach. In: Domingo-Ferrer, J., Torra, V. (eds.) PSD 2004. LNCS, vol. 3050, pp. 162–174. Springer, Heidelberg (2004)
27. Aggarwal, G., Feder, T., Kenthapadi, K., Khuller, S., Panigrahy, R., Thomas, D., Zhu, A.: Achieving anonymity via clustering. In: Proc. of the 25th ACM Symp. on Principles of Database Systems (PODS 2006) (2006)

28. Li, J., Wong, R.C.W., Fu, A.W.C., Pei, J.: Achieving k-anonymity by clustering in attribute hierarchical structures. In: Tjoa, A.M., Trujillo, J. (eds.) DaWaK 2006. LNCS, vol. 4081, pp. 405–416. Springer, Heidelberg (2006)

29. Byun, J.W., Kamra, A., Bertino, E., Li, N.: Efficient k-anonymization using clustering techniques. In: Kotagiri, R., Radha Krishna, P., Mohania, M., Nantajeewarawat, E. (eds.) DASFAA 2007. LNCS, vol. 4443, pp. 188–200. Springer, Heidelberg (2007)

30. Aggarwal, C.C., Yu, P.S.: A condensation approach to privacy preserving data mining. In: Bertino, E., Christodoulakis, S., Plexousakis, D., Christophides, V., Koubarakis, M., Böhm, K., Ferrari, E. (eds.) EDBT 2004. LNCS, vol. 2992, pp. 183–199. Springer, Heidelberg (2004)

31. Aggarwal, C.C., Yu, P.S.: On anonymization of string data. In: Proc. of the 2007 SIAM Int. Conf. on Data Mining (2007)

32. Aggarwal, C.C.: On k-anonymity and the curse of dimensionality. In: Proc. of the 31st Int. Conf. on Very Large Databases (VLDB 2005) (2005)

33. Atzori, M.: Weak k-anonymity: A low-distortion model for protecting privacy. In: Katsikas, S.K., López, J., Backes, M., Gritzalis, S., Preneel, B. (eds.) ISC 2006. LNCS, vol. 4176, pp. 60–71. Springer, Heidelberg (2006)

34. Xiao, X., Tao, Y.: Anatomy: Simple and effective privacy preservation. In: Proc. of the 32nd Int. Conf. on Very Large Databases (VLDB 2006) (2006)

35. Kifer, D., Gehrke, J.: Injecting utility into anonymized datasets. In: Proc. of the 2006 ACM SIGMOD Int. Conf. on Management of Data (SIGMOD 2006) (2006)

36. Øhrn, A., Ohno-Machado, L.: Using boolean reasoning to anonymize databases. Artificial Intelligence in Medicine 15(3), 235–254 (1999)

37. Li, N., Li, T., Venkatasubramanian, S.: t-closeness: Privacy beyond k-anonymity and l-diversity. In: Proc. of the 23rd IEEE Int. Conf. on Data Engineering (ICDE 2007) (2007)

38. Bayardo, R., Agrawal, R.: Data privacy through optimal k-anonymity. In: Proc. of the 21st IEEE Int. Conf. on Data Engineering (ICDE 2005) (2005)

39. LeFevre, K., DeWitt, D.J., Ramakrishnan, R.: Mondrian multidimensional k-anonymity. In: Proc. of the 22nd IEEE Int. Conf. on Data Engineering (ICDE 2006) (2006)

40. Gruteser, M., Grunwald, D.: Anonymous Usage of Location-Based Services Through Spatial and Temporal Cloaking. In: Proc. of the First Int. Conf. on Mobile Systems, Applications, and Services (MobiSys 2003) (2003)

41. Gedik, B., Liu, L.: Location Privacy in Mobile Systems: A Personalized Anonymization Model. In: Proc. of the 25th Int. Conf. on Distributed Computing Systems (ICDCS 2005) (2005)

42. Domingo-Ferrer, J.: Microaggregation for database and location privacy. In: Etzion, O., Kuflik, T., Motro, A. (eds.) NGITS 2006. LNCS, vol. 4032, pp. 106–116. Springer, Heidelberg (2006)

43. Kido, H., Yanagisawa, Y., Satoh, T.: Protection of Location Privacy using Dummies for Location-based Services. In: Proc. of the 21st IEEE Int. Conf. on Data Engineering (ICDE 2005) (2005)

44. Beresford, A.R., Stajano, F.: Mix Zones: User Privacy in Location-aware Services. In: Proc. of the Second IEEE Conf. on Pervasive Computing and Communications Workshops (PERCOM 2004) (2004)

45. Gruteser, M., Liu, X.: Protecting Privacy in Continuous Location-Tracking Applications. IEEE Security & Privacy Magazine 2(2), 28–34 (2004)

46. Bettini, C., Wang, X.S., Jajodia, S.: Testing complex temporal relationships involving multiple granularities and its application to data mining. In: Proc. of the 15th ACM Symp. on Principles of Database Systems (PODS 1996) (1996)
47. Giannotti, F., Nanni, M., Pinelli, F., Pedreschi, D.: Trajectory pattern mining. In: Proc. of the 13th ACM SIGKDD Int. Conf. on Knowledge Discovery and Data Mining (KDD 2007) (2007)
48. Trajcevski, G., Wolfson, O., Hinrichs, K., Chamberlain, S.: Managing uncertainty in moving objects databases. ACM Trans. Database Syst. 29(3), 463–507 (2004)
49. Clifton, C., Marks, D.: Security and privacy implications of data mining. In: Proc. of the 1996 ACM SIGMOD Int. Conf. on Management of Data (SIGMOD 1996), February 1996, pp. 15–19 (1996)
50. O'Leary, D.E.: Knowledge discovery as a threat to database security. In: Piatetsky-Shapiro, G., Frawley, W.J. (eds.) Knowledge Discovery in Databases, pp. 507–516. AAAI/MIT Press (1991)
51. Agrawal, R., Srikant, R.: Privacy-preserving data mining. In: Proc. of the 2000 ACM SIGMOD Int. Conf. on Management of Data (SIGMOD 2000), pp. 439–450 (2000)
52. Clifton, C., Kantarcioglu, M., Vaidya, J.: Defining privacy for data mining. In: Natural Science Foundation Workshop on Next Generation Data Mining, November 2002, pp. 126–133 (2002)
53. Verykios, V.S., Bertino, E., Fovino, I.N., Provenza, L.P., Saygin, Y., Theodoridis, Y.: State-of-the-art in privacy preserving data mining. ACM SIGMOD Record 33(1), 50–57 (2004)
54. Inan, A., Saygin, Y.: Privacy-preserving spatio-temporal clustering on horizontally partitioned data. In: Tjoa, A.M., Trujillo, J. (eds.) DAWAK 2006. LNCS, vol. 4081, pp. 459–468. Springer, Heidelberg (2006)
55. Abul, O., Atzori, M., Bonchi, F., Giannotti, F.: Hiding sequences. In: Proceedings of the Third ICDE International Workshop on Privacy Data Management (PDM 2007) (2007)
56. Abul, O., Atzori, M., Bonchi, F., Giannotti, F.: Hiding sensitive trajectory patterns. In: ICDM 2007, pp. 693–698 (2007)
57. Bonchi, F., Saygin, Y., Verykios, V.S., Atzori, M., Gkoulalas-Divanis, A., Kaya, S.V., Savas, E.: Privacy in spatiotemporal data mining. In: [1], pp. 297–333

Author Index